Student Solutions Manual
to accompany

BUSINESS MATHEMATICS
A Collegiate Approach

Ninth Edition

Nelda W. Roueche

Virginia H. Graves
Northern Virginia Community College, Alexandria

Michael D. Tuttle
Northwood University

PEARSON
Prentice
Hall

Upper Saddle River, New Jersey
Columbus, Ohio

Senior Acquisitions Editor: Gary Bauer
Editorial Assistant: Jacqueline Knapke
Production Editor: Louise N. Sette
Design Coordinator: Diane Ernsberger
Cover Designer: Thomas Mack
Production Manager: Pat Tonneman
Marketing Coordinator: Leigh Ann Sims

Pearson Education Ltd.
Pearson Education Singapore Pte. Ltd.
Pearson Education Canada, Ltd.
Pearson Education—Japan

Pearson Education Australia Pty. Limited
Pearson Education North Asia Ltd.
Pearson Educación de Mexico, S.A. de C.V.
Pearson Education Malaysia Pte. Ltd.

10 9 8 7 6 5 4 3

ISBN 0-13-114022-1

CONTENTS

Section 1

1. (a) millions   (d) hundred thousands
   (b) tens   (e) ones
   (c) billions

3. (a) Six hundred thirty-three million, five hundred twenty
       thousand, four hundred eighty-one.
   (b) Twenty-five million, five hundred forty-three thousand,
       one hundred twenty-eight.
   (c) One hundred fifty million, two hundred eighty-six
       thousand, four hundred thirteen.
   (d) Six million, forty-six thousand, one hundred twenty-five.
   (e) Eight hundred twelve billion, three hundred forty-four
       million, six hundred one thousand, twenty-two.

Section 2

1.	(a)	20	3.	(a)	22
	(b)	19		(b)	24
	(c)	27		(c)	28
	(d)	25		(d)	37
	(e)	27		(e)	284
	(f)	20		(f)	221
	(g)	19		(g)	213
	(h)	225		(h)	315
	(i)	206		(i)	268

5.	(a)	206	7.	(a)	8
	(b)	262		(b)	664
	(c)	1,412		(c)	1,921
				(d)	323

9. (a) 4,660
   (b) 6,400
   (c) 55,300
   (d) 11,200
   (e) 82,800
   (f) 2,075,000

11. (a) 1,824
    (b) 37,842
    (c) 381,351
    (d) 1,750,060

13. (a) 598
    (b) 746
    (c) 18
    (d) 26
    (e) 291

15. Daily Totals: Dept. Totals:
    Monday $ 2,457.47   Dept. #1 $ 2,404.35
    Tuesday   2,315.53   #2   3,002.09
    Wednesday   2,189.82   #3   2,651.89
    Thursday   2,361.51   #4   1,991.98
    Friday   2,732.17   #5   1,962.61
    Saturday   2,627.90   #6   2,671.48
                $14,684.40 $14,684.40

Section 3

1. (a) $\frac{2}{3}$ (d) $\frac{4}{7}$ (g) $\frac{3}{4}$

    (b) $\frac{7}{3}$ or $2\frac{1}{3}$ (e) $\frac{2}{3}$ (h) $\frac{6}{5}$ or $1\frac{1}{5}$

    (c) $\frac{7}{9}$ (f) $\frac{5}{3}$ or $1\frac{2}{3}$ (i) $\frac{63}{107}$

                                                       (j) $\frac{7}{12}$

3. (a) By common denominator:

    $\frac{24}{56}, \frac{42}{56}, \frac{20}{56}, \frac{28}{56}, \frac{35}{56}, \frac{40}{56}$

    By order of size:

    $\frac{5}{14}, \frac{3}{7}, \frac{1}{2}, \frac{5}{8}, \frac{5}{7}, \frac{3}{4}$

    (b) By common denominator:

    $\frac{42}{72}, \frac{60}{72}, \frac{54}{72}, \frac{56}{72}, \frac{45}{72}, \frac{39}{72}$

    By order of size:

    $\frac{13}{24}, \frac{7}{12}, \frac{5}{8}, \frac{3}{4}, \frac{7}{9}, \frac{5}{6}$

5. (a) $2\frac{5}{6}$ (e) $16\frac{1}{3}$ (i) $5\frac{2}{3}$

  (b) $9\frac{1}{5}$ (f) $3\frac{1}{12}$ (j) $11\frac{1}{5}$

  (c) $8\frac{1}{2}$ (g) $15\frac{1}{3}$ (k) $11\frac{1}{3}$

  (d) $3\frac{3}{7}$ (h) $13\frac{1}{2}$ (l) $9\frac{1}{2}$

7. (a) $\frac{1}{2}$ (c) $\frac{25}{56}$ (e) $\frac{1}{15}$ (g) $18\frac{5}{12}$ (i) $15\frac{2}{7}$

  (b) $\frac{11}{20}$ (d) $\frac{1}{12}$ (f) $11\frac{11}{12}$ (h) $20\frac{5}{6}$ (j) $20\frac{1}{2}$

9. (a) $\frac{1}{40}$ (b) $\frac{7}{32}$ (c) $24\frac{1}{2}$ (d) 28

Section 4

1. (a) thousandths (d) hundreds (g) ten thousandths

  (b) ones (e) hundredths (h) tenths

  (c) tens (f) thousands (i) hundred thousandths

3. (a) 0.4 (j) $0.54\frac{6}{11}$ or $0.54\overline{54}$

  (b) 0.25 (k) $0.30\frac{10}{13}$ or 0.307692

  (c) 0.3 (l) $0.26\frac{12}{13}$ or 0.2692307

  (d) 0.7 (m) 6.0

  (e) 0.32 (n) $40\frac{3}{4}$ or 40.75

  (f) 0.075 (o) 5.4

3. (Continued)

(g) $0.14\frac{2}{7}$ or 0.142857   (p) $5.11\frac{1}{9}$ or $5.11\overline{1}$

(h) $0.44\frac{4}{9}$ or $0.44\overline{4}$ (q) $3.83\frac{1}{3}$ or $3.83\overline{3}$

(i) $0.41\frac{2}{3}$ or $0.416\overline{6}$ (r) 9.428

5. (a) 35.62   (e) 884.52   (i) 280   (m) 545,000

(b) 26.45   (f) 1.6   (j) 0.003651   (n) 0.009945

(c) 5.5663   (g) 265   (k) 52,500

(d) 0.10488   (h) 881.2   (l) 0.00644

7. (a) 46.95   (e) 7.811   (i) 29.0096   (m) 37.697

(b) 3.725   (f) 3.845   (j) 142.344   (n) 48.2

(c) 28.859   (g) 558.822   (k) 521.279   (o) 285.12

(d) 240.316   (h) 12.929   (l) 455.045   (p) 68.894

(q) 4.5788

9. (a) 78   (e) 136   (i) 0.344   (m) 440

(b) 123   (f) 3,270   (j) 32.58   (n) 41

(c) 46   (g) 1.27   (k) 0.78   (o) 321

(d) 68   (h) 0.24   (l) 0.0645   (p) 0.28

Appendix B - Metric & Currency Conversions
Problem Solutions

Section 1

1. (a) <u>10</u> meters   (g) 1 <u>kilometer</u>

    (b) 1 <u>hectoliter</u>   (h) <u>0.001</u> liter

    (c) 1 <u>milligram</u>   (i) <u>1000</u> decigrams

    (d) <u>0.1</u> liter   (j) 1 <u>milliliter</u>

    (e) <u>100</u> centimeters   (k) <u>100</u> decameters

    (f) 1 <u>gram</u>   (l) 1 <u>decagram</u>

3. (a) <u>360</u> liters   (f) <u>1.88</u> hectoliters

    (b) <u>4.5</u> meters   (g) <u>300</u> decimeters

    (c) <u>2.150</u> grams   (h) <u>1.2</u> kilograms

    (d) <u>800</u> meters   (i) <u>29</u> milligrams

    (e) <u>425</u> centigrams   (j) <u>0.76</u> hectometers

5. (a) 6 meters = <u>6.54</u> yards
   6(1.09 yds.) = 6.554

    (b) 50 kilograms = <u>110</u> pounds
   50(2.20 pounds) = 110

    (c) 10 liters = <u>10.6</u> quarts
   10(1.06 quarts) = 10.6

    (d) 80 kilometers = <u>49.68</u> miles
   80(0.621 miles) = 49.68

    (e) 20 centimeters = <u>7.88</u> inches
   20(0.394 inches) = 7.88

    (f) 0.5 liter = <u>16.9</u> ounces
   0.5(33.8 ounces) = 16.9

    (g) 5 meters = <u>16.4</u> feet
   5(3.28 feet) = 16.4

    (h) 400 grams = <u>14</u> ounces
   400(0.035 ounces) = 14

7. (a) 8 feet = <u>2.44</u> meters
   8(0.305 meters) = 2.44

   (b) 100 pounds = <u>45.4</u> kilograms
   100(0.454 kilograms) = 45.4

   (c) 4 quarts = <u>3.784</u> liters
   4(0.946 liters) = 3.784

   (d) 20 pints = <u>9.46</u> liters
   20(0.473 liters) = 9.46

   (e) 80 miles = <u>128.8</u> kilometers
   80(1.61 kilometers) = 128.8

   (f) 6 inches = <u>15.24</u> centimeters
   6(2.54 centimeters) = 15.24

   (g) 9 ounces = <u>255.6</u> grams
   9(28.4 grams) = 255.6

   (h) 5 gallons = <u>18.95</u> liters
   5(3.79 liters) = 18.95

9. (a) $C = \dfrac{5}{9}(F - 32)$

   $= \dfrac{5}{9}(77 - 32)$

   $= \dfrac{5}{9}(45)$

   $C = 25°$

   (b) $C = \dfrac{5}{9}(F - 32)$

   $= \dfrac{5}{9}(95 - 32)$

   $= \dfrac{5}{9}(63)$

   $C = 35°$

   (c) $C = \dfrac{5}{9}(F - 32)$

   $= \dfrac{5}{9}(59 - 32)$

   $= \dfrac{5}{9}(27)$

   $C = 15°$

   (d) $C = \dfrac{5}{9}(F - 32)$

   $= \dfrac{5}{9}(14 - 32)$

   $= \dfrac{5}{9}(-18)$

   $C = -10°$

9. (Continued)

(e) $F = \dfrac{9}{5}C + 32$

   $ = \dfrac{9}{5}(20) + 32$

   $ = 36 + 32$

   $F = 68°$

(f) $F = \dfrac{9}{5}C + 32$

   $ = \dfrac{9}{5}(80) + 32$

   $ = 144 + 32$

   $F = 176°$

(g) $F = \dfrac{9}{5}C + 32$

   $ = \dfrac{9}{5}(55) + 32$

   $ = 99 + 32$

   $F = 131°$

(h) $F = \dfrac{9}{5}C + 32$

   $ = \dfrac{9}{5}(-5) + 32$

   $ = -9 + 32$

   $F = 23°$

11. (a) 120 inches = <u>304.8</u> centimeters
    120(2.54 centimeters) = 304.8

    (b) 250 centimeters = <u>98.5</u> inches
    250(0.394 inches) = 98.5

    (c) American car: 304.8 - 250 = 54.8 centimeters

13. (a) 5 feet 6 inches = 5 × 12 + 6 = 66 inches
    66 inches = <u>167.64</u> centimeters
    66(2.54 centimeters) = 167.64

    (b) 120 pounds = <u>54.48</u> kilograms
    120(0.454 kilograms) = 54.48

15. 3000 miles = <u>4830</u> kilometers
    3000(1.61 kilometers) = 4830
    25000 miles = <u>40250</u> kilometers
    25000(1.61 kilometers) = 40250

17. (a) 20 grams = <u>0.7</u> ounces
    20(0.035 ounces) = 0.7

    (b) 1.75 kilograms = <u>3.85</u> pounds
    1.75(2.2 pounds) = 3.85

19. (a) $95° F = \underline{35° C}$

$$C = \frac{5}{9}(F - 32)$$

$$= \frac{5}{9}(95 - 32)$$

$$= \frac{5}{9}(63)$$

$$C = 35°$$

(b) $30° C = \underline{86° F}$

$$F = \frac{9}{5}C + 32$$

$$= \frac{9}{5}(30) + 32$$

$$= 54 + 32$$

$$F = 86°$$

(c) New York: $35° C - 30° C = 5° C$

Section 2

		U.S. ($)	Euro	Japanese (¥)	Mexican ($)
1.	(a)	$ 10	9.01	¥ 1,083	$ 94.82
	(b)	600	540.36	64,962	5,689.20
	(c)	1,800	1,621.08	194,886	17,067.60
3.	(a)	$3,122.44	2812	X	X
	(b)	494.59	X	¥ 53550	X
	(c)	357.52	X	X	$ 3390

1. (a) $10 × 0.9006 = 9.01
$10 × 108.27 = ¥1,083
$10 × 9.482 = $94.82

(b) $600 × 0.9006 = 540.36
$600 × 108.27 = ¥64,962
$600 × 9.482 = $5,689.20

(c) $1,800 × 0.9006 = 1,621.08
$1,800 × 108.27 = ¥194,886
$1,800 × 9.482 = $17,067.60

3. (a) 2812 × 1.1104   = $3,122.44
(b) ¥53550 × 0.009236 = $494.59
(c) $3,390 × 0.105463 = 357.519 = $357.52

5. $8,000 × 1.3831 = C$11,064.80

7. $80 × 0.9006 = 72.05

9. £50.144 × 1.6108 = 80.7719 = $80.77

11. 165 × 1.1104 = $183

13. (a) $3.99 × 0.9006 = 3.59 Euros
    (b) $3.99 × 108.27 = ¥432
    (c) $3.99 × 1.3831 = C$5.52

Section 1

1. (a)
```
      83
  ×   30
   1,530
```
(b)
```
     256
  ×   50
  12,800
```
(c)
```
    18 3
  ×  600
  109,800
```

(d)
```
      44
  × 1300
   13 2
   44
   57,200
```
(e)
```
     118
  × 3600
   70 8
   354
   424,800
```
(f)
```
    1641
  ×  302
   3 282
   492 3
   495,582
```

(g)
```
     1765
  ×   405
   8 825
   706 0
   714,825
```
(h)
```
       13202
  ×     2009
   118 818
   26 404
   26,522,818
```
(i)
```
      621
  × 1070
   43 47
   621
   664,470
```

(j)
```
        34
  ×  6.4 1/2
     1 7
    13 6
    204
    219.3
```
(k)
```
      5.2
  × 18 1/4
     1 3
    41 6
    52
    94.9
```
(l)
```
     0.48
  ×  28 1/3
      16
     2 24
    11 2
    13.60
```

(m)
```
      5.4
  ×   32 1/6
      9
     10 8
     162
     173.7
```
(n)
```
      2.4
  × 3.5 3/4
      18
     1 20
     7 2
     8.58
```

(o) $5^4 = 5 \times 5 \times 5 \times 5 = 625$

(p) $6^3 = 6 \times 6 \times 6 = 216$

(q) $7^3 = 7 \times 7 \times 7 = 343$

(r) $2.01^2 = 2.01 \times 2.01 = 4.0401$

3. (a) $\dfrac{48}{\frac{6}{7}} = \dfrac{48}{1} \div \dfrac{6}{7} = \dfrac{48}{1} \times \dfrac{7}{6} = 56$

   (b) $\dfrac{72}{4\frac{1}{2}} = \dfrac{72}{1} \div \dfrac{9}{2} = \dfrac{72}{1} \times \dfrac{2}{9} = 16$

   (c) $\dfrac{40}{0.5} = \dfrac{400}{5} = 80$ (d) $\dfrac{48.24}{0.12} = \dfrac{4824}{12} = 402$

   (e) $\dfrac{1.36}{0.8} = \dfrac{136}{80} = 1.7$ (f) Net sales: $ 66,708
                                                        Cost of goods sold: −68,134
                                                        Loss: <$ 1,426>

   (g) Travel allowance: $ 300
        Travel expenses: −475
        Deficit: <$ 175>

   (h) Net income: $ 82,500
        Partner's salaries: −100,000
        Deficit: <$ 17,500>

   (i) Escrow for taxes: $ 2,575
        Taxes assessed: −3,100
        Balance: <$ 525>

5. (a) $\left(1 + \dfrac{1}{10} \cdot \dfrac{1}{4}\right) = \left(1 + \dfrac{1}{40}\right) = \dfrac{41}{40} = 1\dfrac{1}{40}$

   (b) $\left(1 + \dfrac{7}{100} \cdot \dfrac{1}{3}\right) = \left(1 + \dfrac{7}{300}\right) = \dfrac{307}{300}$ or $1\dfrac{7}{300}$

   (c) $\left(1 - \dfrac{4}{100} \cdot \dfrac{1}{8}\right) = \left(1 - \dfrac{1}{200}\right) = \dfrac{199}{200}$

   (d) $\left(1 - \dfrac{4}{100} \cdot \dfrac{1}{2}\right) = \left(1 - \dfrac{2}{100}\right) = \dfrac{98}{100} = \dfrac{49}{50}$

   (e) $5,000\left(1 + \dfrac{18}{100} \cdot \dfrac{4}{9}\right) = 5,000\left(1 + \dfrac{2}{25}\right) = 5,000\left(\dfrac{27}{25}\right) = 5,400$

-11-

5. (Continued)

(f) $1,500\left(1 + \dfrac{28}{100} \cdot \dfrac{2}{7}\right) = 1,500\left(1 + \dfrac{4}{50}\right) = 1,500\left(\dfrac{54}{50}\right) = 1,620$

(g) $1,000\left(1 - \dfrac{6}{100} \cdot \dfrac{2}{5}\right) = 1,000\left(1 - \dfrac{6}{250}\right) = 1,000\left(\dfrac{244}{250}\right) = 976$

(h) $500\left(1 - \dfrac{12}{100}g\right) = 500(1) - 500\left(\dfrac{12}{100}g\right) = 500 - 60g$

(i) $300\left(1 + \dfrac{9}{100}b\right) = 300(1) + 300\left(\dfrac{9}{100}b\right) = 300 + 27b$

(j) $j(kl + m) = j(kl) + j(m) = jkl + jm$

(k) $r(1 - zx) = r(1) - r(zx) = r - rzx$

7. (a)   43.25   =   43.3   (b)   8.942   =   8.94
          56.659 =   56.7   26.4453 =   26.45
      1,680.952 = 1,681.0   121.452   = 121.45

(c) 499.9999 = 500.000   (d)   5.08473 = 5.1; 5.08; 5.085
      0.5664   =   0.566   23.67521 = 23.7; 23.68; 23.675
    337.00894 = 337.009

9. (a) 14.2 × 12.35 = 175.4; 175

(b) 4.56 × 7.3 = 33.3; 33

(c) 5.8 × 7.83 = 45.4; 45

(d) 1.111 × 3.85 = 4.28; 4.28

11. (a) $300 × 1.91301845
         $300 × 2 = 600.00 (5 digits); 5 + 1 = 6 digits
         $300 × 1.91302 = $573.91

(b) $500 × 1.52161826
     $500 × 1.5 = 750.00 (5 digits); 5 + 1 = 6 digits
     $500 × 1.52162 = $760.81

11. (Continued)

   (c) $4,000 × 0.37440925
        $4,000 × 0.4 = 1,600.00 (6 digits); 6 + 1 = 7 digits
        $4,000 × 0.3744093 = $1,497.6372 = $1,497.64

   (d) $20 × 19.08162643
        $20 × 19 = 360.00 (5 digits); 5 + 1 = 6 digits
        $20 × 19.0816 = $381.63

Section 2

1. (a) 2,620 - 48 + 188 - 251 = 2,509

   (b) 964 - 410 + 17 = 571

   (c) 41 ÷ 2 × 6 = 123   (d) 12 ÷ 8 × 46 = 69

   (e) $\frac{3}{8}$ × 8,400 = 3,150   (f) $\frac{4}{9}$(108) = 48

   (g) $\frac{2}{7}$(420) = 120   (h) $\frac{42}{3}$ × 7 = 98

3. (a) 471 × 15% = 70.65   (b) 582 × 30 × 5% = 873

   (c) 609 × 6.2% = 37.758   (d) 1,500 × 90% × 60% = 810

5. (a) 0.1225 M+; 32 × MR = 3.92; 60 × MR = 7.35; 180 × MR = 22.05

   (b) 0.14 M+; 12 × MR = 1.68; 25 × MR = 3.5; 1,400 × MR = 196

   (c) $\frac{41 + 785}{1 - 0.80}$: 1 - 0.80 M+ → 0.2; 41 + 785 ÷ MR = 4,130

   (d) $\frac{461 - 56.42 - 21.26}{12.34 + 0.61}$: 12.34 + 0.61 M+ → 12.95;
        461 - 56.42 - 21.26 ÷ MR = 29.6

   (e) 4,000(1 + 0.08 × 24): 1 M+; 0.08 × 24 M+ → 1.92;
        MR → 292 × 4,000 = 11,680

5. (Continued)

(f) $66\left(1 - \dfrac{3}{4} \times 9\%\right)$: 1 M+; 0.75 × 9% M- → 0.0675;

  MR → 0.9325 × 66 = 61.545

(g) $720\left(1 + \dfrac{5}{8} \times 5\%\right)$: 1 M+; 0.625 × 9% M+ → 0.03125;

  MR → 1.03125 × 720 = 742.50

(h) $\dfrac{49,350}{47 \times 15\%}$: 47 × 15% M+ → 7.05; 49,350 ÷ MR = 7,000

Section 1

1.
$$y + 14 = 65$$
$$y + \cancel{14} - \cancel{14} = 65 - 14$$
$$y = 51$$

3.
$$x - 12 = 56$$
$$x - \cancel{12} + \cancel{12} = 56 + 12$$
$$x = 68$$

5.
$$8z = 96$$
$$\frac{\cancel{8}z}{\cancel{8}} = \frac{96}{8}$$
$$z = 12$$

7.
$$12b = 108$$
$$\frac{\cancel{12}b}{\cancel{12}} = \frac{108}{12}$$
$$b = 9$$

9.
$$4x + 7 = 23$$
$$4x + \cancel{7} - \cancel{7} = 23 - 7$$
$$4x = 16$$
$$\frac{\cancel{4}x}{\cancel{4}} = \frac{16}{4}$$
$$x = 4$$

11.
$$7c + 48 = 125$$
$$7c + \cancel{48} - \cancel{48} = 125 - 48$$
$$7c = 77$$
$$\frac{\cancel{7}c}{\cancel{7}} - \frac{77}{7}$$
$$c = 11$$

13.
$$14s + 10 = 80$$
$$14s + \cancel{10} - \cancel{10} = 80 - 10$$
$$14s = 70$$
$$\frac{\cancel{14}s}{\cancel{14}} = \frac{70}{14}$$
$$s = 5$$

15.
$$8z - 8 = 2z + 52$$
$$8z - \cancel{8} + \cancel{8} = 2z + 52 + 8$$
$$8z = 2z + 60$$
$$8z - 2z = \cancel{2z} - \cancel{2z} + 60$$
$$6z = 60$$
$$\frac{\cancel{6}z}{\cancel{6}} = \frac{60}{6}$$
$$z = 10$$

17. $4t + 10 = 38$

$4t + \cancel{10} - \cancel{10} = 38 - 10$

$4t = 28$

$\dfrac{\cancel{4}t}{\cancel{4}} = \dfrac{28}{4}$

$t = 7$

19. $6(y + 3) = y + 128$

$6y + 18 = y + 128$

$6y + \cancel{18} - \cancel{18} = y + 128 - 18$

$6y = y + 110$

$6y - y = \cancel{y} - \cancel{y} + 110$

$5y = 110$

$\dfrac{\cancel{5}y}{\cancel{5}} = \dfrac{110}{5}$

$y = 22$

21. $8f + 22 = f + 92$

$8f + \cancel{22} - \cancel{22} = f + 92 - 22$

$8f = f + 70$

$8f - f = \cancel{f} - \cancel{f} + 70$

$7f = 70$

$\dfrac{\cancel{7}f}{\cancel{7}} = \dfrac{70}{7}$

$f = 10$

23. $15x - 6 = 7x + 26$

$15x - \cancel{6} + \cancel{6} = 7x + 26 + 6$

$15x = 7x + 32$

$15x - 7x = \cancel{7x} - \cancel{7x} + 32$

$8x = 32$

$\dfrac{\cancel{8}x}{\cancel{8}} = \dfrac{32}{8}$

$x = 4$

25. $2h - 10 = 32 - 4h$

$2h - \cancel{10} + \cancel{10} = 32 + 10 - 4h$

$2h = 42 - 4h$

$2h + 4h = 42 - \cancel{4h} + \cancel{4h}$

$6h = 42$

$\dfrac{\cancel{6}h}{\cancel{6}} = \dfrac{42}{6}$

$h = 7$

27. $6(x - 2) = 4x + 24$

$6x - 12 = 4x + 24$

$6x - \cancel{12} + \cancel{12} = 4x + 24 + 12$

$6x = 4x + 36$

$6x - 4x = \cancel{4x} - \cancel{4x} + 36$

$2x = 36$

$\dfrac{\cancel{2}x}{\cancel{2}} = \dfrac{36}{2}$

$x = 18$

29. $\dfrac{3a}{8} = 12$

$\dfrac{\cancel{8}}{\cancel{3}} \cdot \dfrac{\cancel{3}a}{\cancel{8}} = 12 \cdot \dfrac{8}{3}$

$a = 4 \cdot 8$

$a = 32$

31. $\dfrac{r}{5} - 4 = 31$

$\dfrac{r}{5} - \cancel{4} + \cancel{4} = 31 + 4$

$\dfrac{r}{5} = 35$

$\cancel{5} \cdot \dfrac{r}{\cancel{5}} = 35 \cdot 5$

$r = 175$

33.
$$\frac{3x}{5} + 8 = 98$$

$$\frac{3x}{5} + 8 - 8 = 98 - 8$$

$$\frac{3x}{5} = 90$$

$$\frac{5}{3} \cdot \frac{3x}{5} = \frac{5}{3} \cdot 90$$

$$x = 150$$

35.
$$47 = \frac{3d}{4} - 22$$

$$47 + 22 = \frac{3d}{4} - 22 + 22$$

$$69 = \frac{3d}{4}$$

$$\frac{4}{3} \cdot 69 = \frac{3d}{4} \cdot \frac{4}{3}$$

$$92 = d$$

Section 2

1. (a) $6x$

    (b) $a + o$

    (c) $n + 10$

    (d) $n - 18$

    (e) $\frac{2}{3}c$

    (f) $\frac{1}{4}p + 5$

    (g) $2(r + s)$

    (h) $g = h - \$4$ (i) $d = 2(a + b)$

    (j) $b = 8.5f$

    (k) $m = \frac{1}{3}n - 9$

    (l) $\$10b$

3. What number decreased by 14 yields 56?

    $$n - 14 = 56$$
    $$n - 14 + 14 = 56 + 14$$
    $$n = 70$$

5. Corner Market charges $2.00 less then Wilson's Mart.

    $$12 = W - \$2.00$$
    $$12 + 2.00 = W - 2.00 + 2.00$$
    $$14 = W$$

7. Scott Shope charges $27 more than Garcia Co.

    $$77 = G + 27$$
    $$77 - 27 = G + 27 - 27$$
    $$50 = G$$

9. $\frac{2}{3}$ of sales were charge sales.

    $$\frac{2}{3}s = c$$
    $$\frac{2}{3}(3,000) = c$$
    $$2,000 = c$$

11. $\frac{1}{5}$ of sales were new customers.

$$\frac{1}{5}s = n$$

$$\frac{1}{5}s = 600$$

$$\frac{5}{1} \cdot \frac{1}{5}s = 600 \cdot 5$$

$$s = 3,000$$

13. 8 less than $\frac{1}{4}$ of employees took no sick leave.

$$\frac{1}{4}e - 8 = L$$

$$\frac{1}{4}e - 8 = 50$$

$$\frac{1}{4}e - 8 + 8 = 50 + 8$$

$$\frac{1}{4}e = 58$$

$$\frac{4}{1} \cdot \frac{1}{4}e = 58 \cdot 4$$

$$e = 232$$

15. February utility expenses were 2.5 times March utility expenses.

$$F = 2.5M$$

$$180 = 2.5M$$

$$\frac{180}{2.5} = \frac{2.5}{2.5}M$$

$$72 = M$$

17. 2,000 hours times rate per hour equals $480.

$$\$480 = 2,000x$$

$$\frac{\$480}{2,000} = \frac{2,000}{2,000}x$$

$$\$.24 = x$$

19. Manager salaries are 1.8 times staff salaries. Total salaries are $420,000.

$$m = 1.8s$$

$$\$420,000 = 1.8s + s$$

$$420,000 = 2.8s$$

$$\frac{420,000}{2.8} = \frac{2.8}{2.8}s$$

$$\$150,000 = s$$

$$1.8(150,000) = m$$

$$\$270,000 = m$$

21. Cellular phones and pagers equal 35 items.
    Cellular phones and pagers totaled $1,900.

$$c + p = 35$$
$$c = 35 - p$$

$$\$50(35 - p) + \$60p = \$1,900$$
$$1,750 - 50p + 60p = 1,900$$
$$1{,}750 - 1{,}750 - 50p + 60p = 1,900 - 1,750$$
$$10p = 1,900 - 1,750$$
$$10p = 150$$
$$p = 15$$

$$c = 35 - p$$
$$c = 35 - 15$$
$$c = 20$$

23. Mortgage, taxes, and insurance equals $1,225.
    Mortgage is 9 times insurance. Taxes are $15
    more than insurance.

$$m + i + t = \$1,225$$
$$(9i) + (i) + (i + 15) = 1,225$$
$$11i + 15 = 1,225$$
$$11i = 1,225 - 15$$
$$\frac{11i}{11} = \frac{1,210}{11}$$
$$i = \$110$$

$$m = 9i$$
$$m = 9(\$110) = \$990$$

$$t = i + \$15$$
$$t = \$110 + \$15 = \$125$$

25. Brand A sold 4 times
     brands B and C together.

$$A = 4(B + C)$$
$$64 = 4(7 + C)$$
$$64 = 28 + 4C$$
$$64 - 28 = \cancel{28} - \cancel{28} + 4C$$
$$36 = 4C$$
$$\frac{36}{4} = \frac{\cancel{4}}{\cancel{4}}C$$
$$9 = C$$

27. Jacket, slacks, and flannel shirt cost $209.
     Jacket cost 2.5 times slacks. Shirt cost $7
     less than slacks.

$$j + s + f = \$209$$
$$j = 2.5s$$
$$f = s - \$7$$

$$(2.5s) + (s) + (s - 7) = 209$$
$$4.5s - 7 = 209$$
$$4.5s = 209 + 7$$
$$4.5s = 216$$
$$\frac{\cancel{4.5}s}{\cancel{4.5}} = \frac{216}{4.5}$$
$$s = \$48$$

$$j = 2.5(\$48) = \$120$$
$$f = \$48 - \$7 = \$41$$

29. Cotton shorts plus lycra shorts equal 54.
    Cotton plus lycra totaled $900.

$$c + l = 54$$
$$c = 54 - l$$

$$\$15(54 - l) + \$18l = \$900$$
$$810 - 15l + 18l = 900$$
$$\cancel{810} - \cancel{810} - 15l + 18l = 900 - 810$$
$$3l = 900 - 810$$
$$3l = 90$$
$$\frac{\cancel{3}l}{\cancel{3}} = \frac{90}{3}$$
$$l = 30$$

$$c = 54 - 30 = 24$$

31.
    Canvas plus leather equaled 26 pairs.
    Canvas plus leather totaled $846.

$$C + L = 26$$
$$C = 26 - L$$

$$\$30(26 - L) + \$36L = 846$$
$$780 - 30L + 36L = 846$$
$$780 + 6L = 846$$
$$\cancel{780} - \cancel{780} + 6L = 846 - 780$$
$$6L = 66$$
$$\frac{\cancel{6}}{\cancel{6}}L = \frac{66}{6}$$
$$L = 11$$

$$C = 26 - L$$
$$C = 26 - 11$$
$$C = 15$$

33.   Friday plus Saturday equaled 275 tickets.
      Friday plus Saturday totaled $7,050.

$$f + s = 275$$
$$s = 275 - f$$

$$\$30(f) + \$22(275 - f) = \$7,050$$
$$30f + 6,050 - 22f = 7,050$$
$$8f + 6,050 = 7,050$$
$$8f = 7,050 - 6,050$$
$$8f = 1,000$$
$$\frac{\cancel{8}f}{\cancel{8}} = \frac{1,000}{8}$$
$$f = 125$$

$$s = 275 - 125 = 150$$

35. Desk lamps and floor lamps equaled 200.
     Desk lamps and floor lamps totaled $5,165.

$$d + f = 200$$
$$d = 200 - f$$

$$\$19(200 - f) + \$40(f) = \$5,165$$
$$3,800 - 19f + 40f = 5,165$$
$$21f = 5,165 - 3,800$$
$$21f = 1,365$$
$$\frac{\cancel{21}f}{\cancel{21}} = \frac{1,365}{21}$$
$$f = 65$$

$$d = 200 - 65 = 135$$

Section 3

1. (a) 4 to 1; 4:1; $\dfrac{4}{1}$ (b) 1 to 3; 1:3; $\dfrac{1}{3}$

   (c) 2 to 5; 2:5; $\dfrac{2}{5}$ (d) 2 to 9; 2:9; $\dfrac{2}{9}$

   (e) 9 to 4; 9:4; $\dfrac{9}{4}$ or

   2.25 to 1; 2.25:1; $\dfrac{2.25}{1}$

3. (a) $\dfrac{c}{14} = \dfrac{2}{7}$ (b) $\dfrac{3}{8} = \dfrac{r}{16}$

   $\cancel{14}\left(\dfrac{c}{\cancel{14}}\right) = \left(\dfrac{2}{7}\right)14$ $16\left(\dfrac{3}{8}\right) = \cancel{16}\left(\dfrac{r}{\cancel{16}}\right)$

   $c = 4$ $6 = r$

   (c) $\dfrac{3}{g} = \dfrac{9}{24}$ (d) $\dfrac{15}{10} = \dfrac{9}{z}$

   $3(24) = 9(g)$ $15(z) = 9(10)$

   $72 = 9g$ $15z = 90$

   $\dfrac{72}{9} = \dfrac{\cancel{9}g}{\cancel{9}}$ $\dfrac{\cancel{15}z}{\cancel{15}} = \dfrac{90}{15}$

   $8 = g$ $z = 6$

5. $\dfrac{\text{Carlita}}{\text{Susanne}} = \dfrac{3}{1}$ 7. $\dfrac{\text{Experience}}{\text{Nonexperience}} = \dfrac{18}{5}$

9. $\dfrac{\text{Operating expenses}}{\text{Net sales}} = \dfrac{21}{100}$

   $\dfrac{21}{100} = \dfrac{x}{120,000}$

   $120,000\left(\dfrac{21}{100}\right) = \left(\dfrac{x}{\cancel{120,000}}\right)\cancel{120,000}$

   $25,200 = x$

11. $\dfrac{\text{Annuity}}{\text{Gross wages}} = \dfrac{5}{100}$

$$\dfrac{5}{100} = \dfrac{33}{W}$$

$$5(W) = 33(100)$$

$$5W = 3{,}300$$

$$\dfrac{5W}{5} = \dfrac{3{,}300}{5}$$

$$W = \$660$$

13. $\dfrac{\text{Men}}{\text{Women}} = \dfrac{1}{7}$

$$\dfrac{1}{7} = \dfrac{6}{W}$$

$$(1)W = 7(6)$$

$$W = 42$$

15. $\dfrac{\text{Bus. Admin.}}{\text{Marketing}} = \dfrac{3}{2}$

$$\dfrac{3}{2} = \dfrac{240}{M}$$

$$3(M) = 2(240)$$

$$3M = 480$$

$$\dfrac{3M}{3} = \dfrac{480}{3}$$

$$M = 160$$

17. $\dfrac{\text{Team}}{\text{Individual}} = \dfrac{7}{2}$

$$\dfrac{7}{2} = \dfrac{T}{210}$$

$$2(T) = 7(210)$$

$$2T = 1470$$

$$\dfrac{2T}{2} = \dfrac{1470}{2}$$

$$T = 735$$

19. $\dfrac{\text{Brand } X}{\text{Brand } Y} = \dfrac{5}{1}$

$$\dfrac{5}{1} = \dfrac{70}{Y}$$

$$5(Y) = 70(1)$$

$$5Y = 70$$

$$\dfrac{5Y}{5} = \dfrac{70}{5}$$

$$Y = 14$$

21. $\dfrac{\text{Words}}{\text{Minutes}} = \dfrac{900}{30}$

$$\dfrac{900}{30} = \dfrac{W}{210}$$

$$900(210) = 30(W)$$

$$189{,}000 = 30(W)$$

$$\dfrac{189{,}000}{30} = \dfrac{30W}{30}$$

$$6{,}300 = W$$

23. $\dfrac{\text{Boxes}}{\text{Minutes}} = \dfrac{15}{40}$

$$\dfrac{15}{40} = \dfrac{B}{240}$$

$$40(B) = 15(240)$$

$$40B = 3{,}600$$

$$\dfrac{40B}{40} = \dfrac{3{,}600}{40}$$

$$B = 90$$

25. $\dfrac{\text{Pages}}{\text{Minutes}} = \dfrac{40}{5}$

$$\dfrac{40}{5} = \dfrac{P}{150}$$

$$5(P) = 40(150)$$

$$5P = 6{,}000$$

$$\dfrac{5P}{5} = \dfrac{6{,}000}{5}$$

$$P = 1{,}200$$

27. $\dfrac{\text{Hours}}{\text{Miles}} = \dfrac{6}{360}$

$\dfrac{6}{360} = \dfrac{H}{660}$

$360(H) = 6(660)$

$360H = 3,960$

$\dfrac{360H}{360} = \dfrac{3,960}{360}$

$H = 11$

Section 1

	(a)	(b)
1.	0.07	$\dfrac{7}{100}$
3.	0.32	$\dfrac{8}{25}$
5.	0.05	$\dfrac{1}{20}$
7.	0.273	$\dfrac{273}{1,000}$
9.	0.0625	$\dfrac{1}{16}$
11.	1.56	$\dfrac{39}{25}$
13.	2.418	$\dfrac{1,209}{500}$
15.	0.008	$\dfrac{1}{125}$
17.	0.0125	$\dfrac{1}{80}$
19.	2.07	$\dfrac{207}{100}$
21.	0.002	$\dfrac{1}{500}$
23.	0.003	$\dfrac{3}{1,000}$
25.	0.0025	$\dfrac{1}{400}$
27.	0.0225	$\dfrac{9}{400}$
29.	0.125	$\dfrac{1}{8}$

31. 4%	33. 40%	35. 67.4%
37. 211%	39. 420%	41. 0.1%
43. 50%	45. 0.5%	47. 310%
49. 3%	51. 50%	53. 0.5%
55. 12.5%	57. 175%	59. 480%

Section 2

1. 8% of 4,000 = __?__

$$(0.08)(4,000) = N$$
$$320 = N$$

3. 40% of 640 = __?__

$$(0.40)(640) = X$$
$$256 = X$$

5. 25% of 78 = __?__

$$(0.25)(78) = X$$
$$19.5 = X$$

7. __?__ $= \frac{3}{5}$% of 600

$$X = (0.006)(600)$$
$$X = 3.6$$

9. __?__ $= 12\frac{1}{2}$% of 560

$$X = (0.125)(560)$$
$$X = 70$$

11. __?__ = 30% of 850

$$N = (0.30)(850)$$
$$N = 255$$

13. 0.25% of 14,000 = __?__

$$(0.0025)(14,000) = A$$
$$35 = A$$

15. __?__% of 90 = 15

$$90X = 15$$
$$X = \frac{15}{90}$$
$$X = 16.7\% \text{ or } 16\frac{2}{3}\%$$

17. $28 = \underline{\ ?\ }$ % of 80

$28 = 80x$

$\dfrac{28}{80} = x$

$\dfrac{7}{20} = x$

$x = 35\%$

19. $5.4 = \underline{\ ?\ }$ % of 210

$5.4 = 210p$

$\dfrac{5.4}{210} = p$

$2.6\% = p$

21. $\underline{\ ?\ }$ % of 18 = 4.5

$18X = 4.5$

$X = \dfrac{4.5}{18}$

$X = 25\%$

23. 15% of $\underline{\ ?\ }$ = 48

$0.15N = 48$

$N = \dfrac{48}{0.15}$

$N = 320$

25. 44% of $\underline{\ ?\ }$ = 110

$0.44N = 110$

$N = \dfrac{110}{0.44}$

$N = 250$

27. $240 = 40\%$ of $\underline{\ ?\ }$

$240 = 0.40A$

$\dfrac{240}{0.40} = A$

$600 = A$

29.

$90 = 33\dfrac{1}{3}\%$ of $\underline{\ ?\ }$

$90 = \dfrac{1}{3}N$

$(3)(90) = N$

$270 = N$

31. 1.25% of $\underline{\ ?\ }$ = 50

$0.0125A = 50$

$A = \dfrac{50}{0.0125}$

$A = 4{,}000$

33. 0.7% of $\underline{\ ?\ }$ = 3.5

$0.007N = 3.5$

$N = \dfrac{3.5}{0.007}$

$N = 500$

35. 0.5% of $\underline{\ ?\ }$ = 7

$0.005A = 7$

$A = \dfrac{7}{0.005}$

$A = 1{,}400$

37. Change = 12 - 5 = 7

 —% of original = change

 —% of 5 = 7

 $5X = 7$

 $X = \dfrac{7}{5}$

 $X = 140\%$

39. Change = 55 - 44 = 11

 —% of original = change

 —% of 55 = 11

 $55X = 11$

 $X = \dfrac{11}{55} = \dfrac{1}{5}$

 $X = 20\%$

41. Change = 200 - 130 = 70

 —% of original = change

 —% of 200 = 70

 $200X = 70$

 $X = \dfrac{70}{200} = \dfrac{7}{20}$

 $X = 35\%$

43. Change = 120 - 108 = 12

 —% of original = change

 —% of 108 = 12

 $108X = 12$

 $X = \dfrac{12}{108} = \dfrac{3}{27}$

 $X = 11.1\%$

45. Change = \$600 - \$597 = \$3

 —% of original = change

 —% of 600 = 3

 $600X = 3$

 $X = \dfrac{3}{600} = \dfrac{1}{200}$

 $X = 0.5\%$

 $\left(\text{or } \dfrac{1}{2}\%\right)$

47. $N + 25\%N = 100$

 $1.25N = 100$

 $N = \dfrac{100}{1.25}$

 $N = 80$

49. $A - 25\%A = 36$

 $A - 0.25A = 36$

 $0.75A = 36$

 $A = \dfrac{36}{0.75}$

 $A = 48$

51. $N + 50\%N = 120$

 $1.5N = 120$

 $N = \dfrac{120}{1.5}$

 $N = 80$

53.

$$A - 33\frac{1}{3}\% A = 30$$

$$\frac{2}{3} A = 30$$

$$A = (30)\frac{3}{2}$$

$$A = 45$$

55. $A + 25\% A = 50$

$$1.25A = 50$$

$$A = \frac{50}{1.25}$$

$$A = 40$$

Section 3

1. Overhead is _?_% of sales?

—— of $21,250 = $8,500

$$21,250X = 8,500$$

$$X = \frac{8,500}{21,250}$$

$$X = 40\%$$

3. _?_% of budget was wages?

——% of $780,000 = $720,000

$$780,000X = 720,000$$

$$X = \frac{720,000}{780,000}$$

$$X = 92.3\%$$

5. 75% of revenue is advertisements.

75% of $1,500,000 = A

$$0.75(1,500,000) = A$$

$$\$1,125,000 = A$$

7. 8% of cost paid in dividends.

8% of $85 = D

$$0.08(85) = D$$

$$\$6.80 = D$$

9. 20% of total income received from investments.

20% of T = $15,300

$$0.20T = 15,300$$

$$T = \frac{15,300}{0.20}$$

$$T = \$76,500$$

11. 63% of total assets represents inventory.

63% of A = $252,000

$$0.63A = 252,000$$

$$A = \frac{252,000}{0.63}$$

$$A = \$400,000$$

13.

35% of collections are fees.

35% of \$186,000 = F

0.35(186,000) = F

65,100 = F

15. 80% of total production costs are raw materials.

80% of total = \$28

$$0.80\,T = 28$$

$$T = \frac{28}{0.80}$$

$$T = \$35$$

17. Change = \$47.15 − \$41.00

= \$6.15

—% of original = change

—% of \$41 = \$6.15

$$41\,X = 6.15$$

$$X = \frac{6.15}{41}$$

$$X = 15\%$$

19. Change = \$84.16 − \$80.00

= \$4.16

—% of original = change

—% of \$80 = \$4.16

$$80\,X = 4.16$$

$$X = \frac{4.16}{80}$$

$$X = 5.2\%$$

21. Change = 15 − 9 = 6

—% of original = change

—% of 15 = 6

$$15\,X = 6$$

$$X = \frac{6}{15}$$

$$X = 40\%$$

23. Change = \$30.00 − \$27.60

= \$2.40

—% of original = change

—% of \$30 = \$2.40

$$30\,X = 2.40$$

$$X = \frac{2.40}{30}$$

$$X = 8\%$$

25.

Change $= \$1,500.00 - \$1,320.00$

$= \$180.00$

___% of original = change

___% of $\$1,500 = \$180.00$

$1,500 X = 180.00$

$X = \dfrac{180.00}{1,500}$

$X = 12\%$

27. What number increased by 40% (of itself) yields 70?

$N + 40\% N = 70$

$1.4 N = 70$

$N = \dfrac{70}{1.4}$

$N = 50$

29. What amount decreased by 20% (of itself) equals $40?

$A - 20\% A = \$40$

$0.80 A = 40$

$A = \dfrac{40}{0.80}$

$A = \$50$

31. 20% of customers were teenagers.

20% of $1,045 = T$

$(0.20)(1,045) = T$

$209 = T$

33. Change $= \$206 - \$200$

$= \$6$

___% of original = change

___% of $\$200 = 6$

$200 X = 6$

$X = \dfrac{6}{200}$

$X = \dfrac{3}{100}$

$X = 3\%$

35. 18% of graduates secured jobs.

18% of $G = 72$

$0.18 G = 72$

$G = \dfrac{72}{0.18}$

$G = 400$

37. Change = $210,000 - $203,700

            = $6,300

  —% of original = change

  —% of $210,000 = $6,300

     $210,000X = 6,300

$$X = \frac{6,300}{210,000}$$

$$X = 3\%$$

39. —% of selling price was gross profit?

  —% of $480 = $192

     $480X = 192$

$$X = \frac{192}{480}$$

$$X = 40\%$$

41. Previous price increased by 8% yields latest price.

$$P + 8\%P = \$114.48$$

$$1.08P = 114.48$$

$$P = \frac{114.48}{1.08}$$

$$P = \$106$$

43. Price reduced by 40% (of itself) yields $2,100.

$$P - 40\%P = \$2,100$$

$$0.60P = 2,100$$

$$P = \frac{2,100}{0.60}$$

$$P = \$3,500$$

45. Change = $271,200 - $253,700

            = $17,500

  —% of original = change

  —% of $253,700 = $17,500

     $253,700X = 17,500$

$$X = \frac{17,500}{253,700}$$

$$X = 6.9\%$$

47. ANSWER WILL VARY.

CHAPTER 4 - BASIC STATISTICS AND GRAPHS
PROBLEM SOLUTIONS

Section 1

1. (a) (1) The arithmetic mean:

$$\frac{10+22+29+22+45+30+43+18+39+28+59+41+42+40+12}{15}$$

$$= \frac{480}{15} = 32$$

(2) The median:

10, 12, 18, 22, 22, 28, 29, <u>30</u>, 39, 40, 41, 42, 43, 45, 59

$\frac{15}{2} = 7+;$ so 8th number = 30

(3) The mode: 22 (occurs twice)

(b) (1) The arithmetic mean:

$$\frac{170 + 154 + 120 + 133 + 142 + 161 + 115 + 121 + 130 + 154}{10}$$

$$= \frac{1,400}{10} = 140$$

(2) The median:

115, 120, 121, 130, 133, 142, 154, 154, 161, 170

$\frac{10}{2} = 5;$ so median is the arithmetic mean of 5th and

6th numbers: $\frac{133 + 142}{2} = 137.5$

(3) The mode: 154 (occurs twice)

3.

Final Grade			Quality Points		Credit Hours
D	1	×	4	=	4
C	2	×	3	=	6
A	4	×	4	=	16
B	<u>3</u>	×	<u>5</u>	=	<u>15</u>
	10		16		41

(a) Average quality points per class: $\frac{10}{4} = 2.50$

3. (Continued)

   (b) Average per credit hour: $\dfrac{41}{16}$ = 2.56

5.

Job	# in Job		Weekly Wages	
Manager	1	×	$1,000 =	$ 1,000
Sales representative	4	×	450 =	1,800
Technician	10	×	550 =	5,500
Office support	10	×	400 =	4,000
	25		$2,400	$12,300

   (a) Mean wage per job: $\dfrac{\$2,400}{4}$ = $600

   (b) Mean wage per employee: $\dfrac{\$12,300}{25}$ = $492

7.

Store	Price		# Sold		
A	$ 85	×	12	=	$1,020
B	135	×	8	=	1,080
C	96	×	22	=	2,112
D	89	×	20	=	1,780
E	112	×	9	=	1,008
	$517		71		$7,000

   (a) Mean price per store: $\dfrac{\$517}{5}$ = $103.40

   (b) Mean price per vacuum sold: $\dfrac{\$7,000}{71}$ = $98.59

9.

Date		Change	Amount of Investment		Months Invested		
January	1	---	$3,500	×	2	=	$7,000
March	1	-$200	3,300	×	5	=	16,500
August	1	+ 400	3,700	×	1	=	3,700
September	1	- 600	3,100	×	4	=	12,400
					12		$39,600

   $\dfrac{\$39,600}{12}$ = $3,300 average investment per month

11. $108,000, $110,000, $130,000, $133,000, $138,000,

    $151,000. $\frac{6}{2} = 3$; so median is the arithmetic mean of the

    third and fourth salaries $= \dfrac{\$130,000 + 133,000}{2} = \$131,500$

13. $79, $79, $\underline{\$82, \$82, \$82}$, $86, $86, $94, $124, $130, $158

    The mode: $82 (occurs 3 times)

15. (a) The mean:

$$48 + 50 + 54 + 60 + 48 + 54 + 56 + 60 +$$

$$\frac{59 + 68 + 52 + 60 + 51 + 60 + 45}{15}$$

$$= \frac{825}{15} = 55 \text{ hours}$$

(b) The median:

45, 48, 48, 50, 51, 52, 54, 54, 56, 59, 60, 60, 60, 60, 68

$\dfrac{15}{2} = 7.5$; so median is the 8th number: 54

(c) The mode: 60 hours (occurs 4 times)

17. (a) The mean:

$$\frac{1630 + 1640 + 1630 + 1620 + 1640 + 1560 + 1640 + 1620}{8}$$

$$= \frac{12,980}{8} = 1,622.5$$

(b) The median:

1560, 1620, 1620, $\underline{1630}$, 1630, 1640, 1640, 1640

$\dfrac{8}{2} = 4$; so median is the arithmetic mean of the 4th and 5th

numbers: $= \dfrac{1630 + 1630}{2} = 1630$

(c) The mode: 1,640 (occurs 3 times)

Section 2

1.

Class Interval	Tally	f	Midpoint	f × Midpoint
60 - 79	///	3	70	210
40 - 59	LHT	5	50	250
20 - 39	LHT ////	9	30	270
0 - 19	///	3	10	30
		20		760

(a) Mean: $\dfrac{760}{20} = 38$

(b) Median: (1) $\dfrac{20}{2} = 10$

   (2) $3 + 7 = 10$

   (3) $\dfrac{7}{9} \times 20 = 15.5555\overline{5} = 15.56$

   (4) $20 + 15.56 = 35.56$

(c) Modal class: 20 - 39

3.

Class Interval	Tally	f	Midpoint	f × Midpoint
$1,000 - 1,099	////	4	$1,050	$ 4,200
900 - 999	////	5	950	4,750
800 - 899	LHT ///	8	850	6,800
700 - 799	LHT /	6	750	4,500
600 - 699	LHT	4	650	2,600
500 - 599	///	3	550	1,650
		30		$24,500

(a) Mean: $\dfrac{\$24,500}{30} = \$816.6\overline{6}$

(b) Median: (1) $\dfrac{30}{2} = 15$

   (2) $3 + 4 + 6 + 2 = 15$

   (3) $\dfrac{2}{8} \times 100 = \$25$

   (4) $\$800 + \$25 = \$825$

3. (Continued)

(c) Modal class: $800 - $899

5.

Class Interval	Tally	f	Midpoint	f × Midpoint
$350 - 449	//	2	$400	$ 800
250 - 349	////	4	300	1,200
150 - 249	LHI ///	8	200	1,600
50 - 149	LHI /	6	100	600
		20		$4,200

(a) Mean: $\dfrac{\$4,200}{20}$ = $210

(b) Median: (1) $\dfrac{20}{2}$ = 10

(2)   6 + 4 = 10

(3) $\dfrac{4}{8}$ × $100 = $50

(4) $150 + $50 = $200

(c) Modal class: $150 - $249

7.

Class Interval	Tally	f	Midpoint	f × Midpoint
21 - 23	/	1	22.5	22.5
18 - 20	//	2	19.5	39.0
15 - 17	//	2	16.5	33.0
12 - 14	////	4	13.5	54.0
9 - 11	/	1	10.5	10.5
6 - 8	LHI ////	9	7.5	67.5
3 - 5	LHI	5	4.5	22.5
0 - 2	//	2	1.5	3.0
		26		252.0

(a) Mean: $\dfrac{252}{26}$ = 9.69

(b) Median: (1) $\dfrac{26}{2}$ = 13

(2)   2 + 5 + 6 = 13

(3) $\dfrac{6}{9}$ × 3 = 2

(4)   6 + 2 = 8

(c) Modal class: 6 - 8

Section 3

1. (a) 50%
   (b) 2.5%
   (c) 34% + 13.5% = 47.5%
   (d) 13.5% + 34% + 34% +
       13.5% + 2.5% = 97.5%
   (e) 16% × 200 = 32
   (f) 84% × 200 = 168

3.

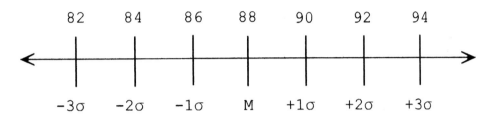

82	84	86	88	90	92	94
-3σ	-2σ	-1σ	M	$+1\sigma$	$+2\sigma$	$+3\sigma$

   (a) 34% + 13.5% + 2.5% = 50%
   (b) 2.5% + 13.5% + 34% + 34% + 13.5% = 97.5%
   (c) 13.5%
   (d) 0%
   (e) 16% of 31 = 4.96 or 5 days
   (f) 97.5% of 31 = 30.2 or 30 days
   (g) 13.5% of 31 = 4.185 or 4 days

5. (a)

Mean	d	d^2	Standard Deviation
8	-5	25	$\sigma = \sqrt{\dfrac{\Sigma d^2}{n}}$
10	-3	9	
12	-1	1	$= \sqrt{\dfrac{80}{5}}$
16	+3	9	
			$= \sqrt{16}$
$\underline{19}$	$\underline{+6}$	$\underline{36}$	
$5\overline{)65} = 13$	0	$\Sigma d^2 = 80$	$\sigma = \pm 4$

(b)

1	5	9	13	17	21	25
-3σ	-2σ	-1σ	M	$+1\sigma$	$+2\sigma$	$+3\sigma$

5. (Continued)

(c) ±1σ: 68% of 5 numbers = 3.4; we have 3: 10, 12, 16.

±2σ: 95% of 5 numbers = 4.75 or 5; all numbers are between 5 and 21.

7. (a)

Mean	d	d^2	Standard Deviation
28	-8	64	$\sigma = \sqrt{\dfrac{\Sigma d^2}{n}}$
34	-2	4	
35	-1	1	$= \sqrt{\dfrac{150}{6}}$
35	-1	1	$= \sqrt{25}$
40	+4	16	$\sigma = \pm 5$

$6\overline{)216} = 36$ (with 44 above) $\dfrac{+8}{0}$ $\Sigma d^2 = \dfrac{64}{150}$

(b)

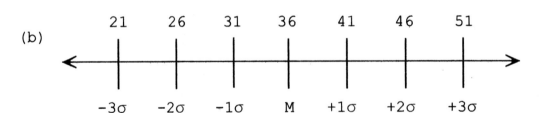

(c) ±1σ: 68% of 6 numbers = 4.08; we have 4: 34, 35, 35, 40.

±2σ: 95% of 6 numbers = 5.7 or 6; we have all 6 numbers between 26 and 46.

9. (a)

Mean	d	d^2	Standard Deviation
31	-9	81	$\sigma = \sqrt{\dfrac{\Sigma d^2}{n}}$
32	-8	64	
35	-5	25	$= \sqrt{\dfrac{288}{8}}$
41	1	1	
44	4	16	$= \sqrt{36}$
44	4	16	$\sigma = \pm 6$
46	6	36	
47	7	49	

$8\overline{)320} = 40 \qquad \overline{0} \qquad \Sigma d^2 = 288$

(b)

```
      22    28    34    40    46    52    58
   ←───┼─────┼─────┼─────┼─────┼─────┼─────┼───→

      -3σ   -2σ   -1σ    M    +1σ   +2σ   +3σ
```

(c) ±1σ: 68% of 8 numbers = 5.44; we have 5: 35, 41, 44, 44, 46.

±2σ: 95% of 8 numbers = 7.6 or 8; we have all 8 numbers between 28 and 52.

11. (a)

Mean	d	d^2	Standard Deviation
23	-7	49	$\sigma = \sqrt{\dfrac{\Sigma d^2}{n}}$
24	-6	36	
25	-5	25	$= \sqrt{\dfrac{200}{8}}$
31	1	1	$= \sqrt{25}$
31	1	1	
34	4	16	$\sigma = \pm 5$
36	6	36	
36	6	36	

$8\overline{)240} = 30 \qquad \overline{0} \qquad \Sigma d^2 = \overline{200}$

(b)

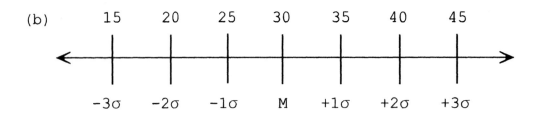

(c) $\pm 1\sigma$: 68% of 8 numbers = 5.44; we have 4: 25, 31, 31, 34.

$\pm 2\sigma$: 95% of 8 numbers = 7.6 or 8; we have all 8 numbers between 20 and 40.

*See note at the end of this section.

13. (a)

Mean	d	d^2	Standard Deviation
14	-6	36	$\sigma = \sqrt{\dfrac{\Sigma d^2}{n}}$
15	-5	25	
16	-4	16	$= \sqrt{\dfrac{160}{10}}$
18	-2	4	
20	0	0	$= \sqrt{16}$
20	0	0	$\sigma = \pm 4$
23	3	9	
23	3	9	
25	5	25	
26	6	36	

$$10\overline{)200} = 20 \qquad \overline{0} \qquad \Sigma d^2 = 160$$

(b)

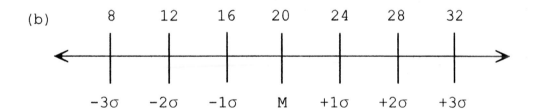

(c) ±1σ: 68% of 10 numbers = 6.8 or 7; we have 6: 16, 18, 20, 20, 23, 23.

±2σ: 95% of 10 numbers = 9.5 or 10; we have all 10 numbers between 12 and 28.

*See note at the end of this section.

Section 4

1. (a) $1,423.00
   (b) $164.50
   (c) $24.14
   (d) New York (176.6)
   (e) New York (114.3)
   (f) Dallas (142.3)
   (g) New York (255.3)
   (h) New York (143.9)
   (i) Atlanta (125.8)

3.

Year	Price	Index (2000 = 100)
2000	$2.50	Base year = 100
2001	2.60	$\frac{2.60}{2.50} \times 100 = 1.04 \times 100 = 104$
2002	2.75	$\frac{2.75}{2.50} \times 100 = 1.10 \times 100 = 110$
2003	2.85	$\frac{2.85}{2.50} \times 100 = 1.14 \times 100 = 114$
2004	2.95	$\frac{2.95}{2.50} \times 100 = 1.18 \times 100 = 118$

5.

Year	Price	Index (1990 = 100)
1990	$0.60	Base year = 100
1995	0.42	$\frac{0.42}{0.60} \times 100 = 0.70 \times 100 = 70$
2000	0.75	$\frac{0.75}{0.60} \times 100 = 1.25 \times 100 = 125$
2005	0.84	$\frac{0.84}{0.60} \times 100 = 1.40 \times 100 = 140$

Section 5

1.

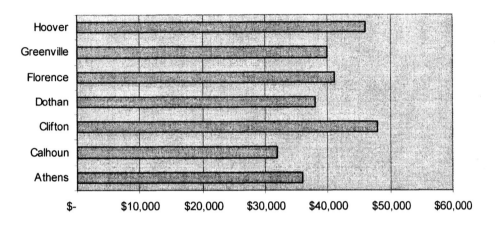

Mean Salaries for Office Managers

3.

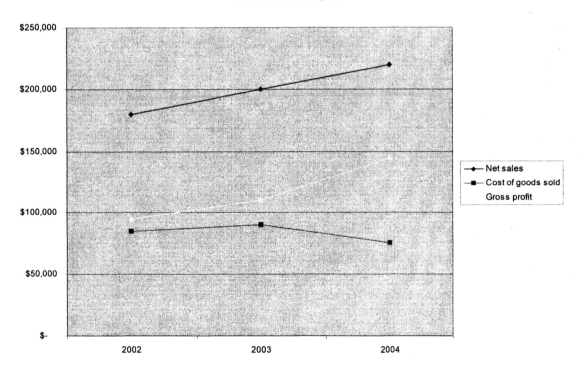

Coleman Interiors, Inc.

5.

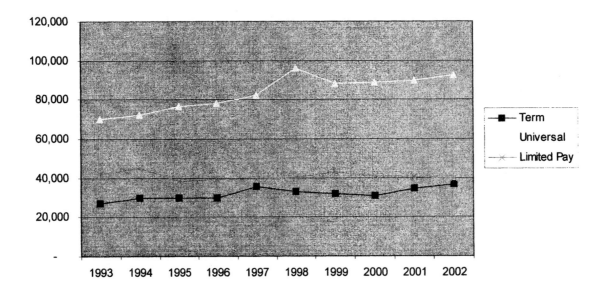

American Insurance Co.

7.

Federal Budget Income Sources

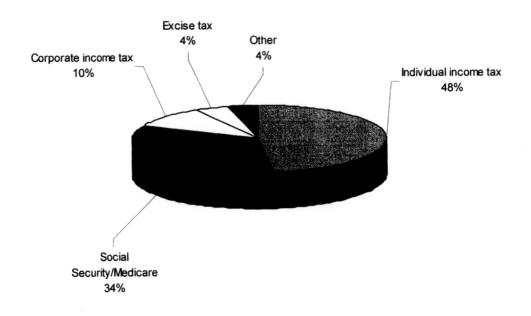

9.

Wadge Industries

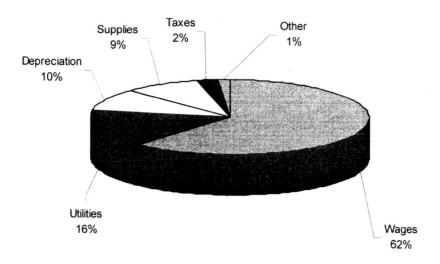

Section 1

1.

	Marked Price	Sales Tax Rate	Sales Tax	Total Price
(a)	$459.00	4%	$18.36	$477.36
(b)	66.00	8	5.28	71.28
(c)	32.50	6	1.95	34.45
(d)	284.00	5	14.20	298.20
(e)	58.00	6	3.48	61.48
(f)	24.00	7	1.68	25.68
(g)	36.00	8	2.88	38.88
(h)	87.00	5	4.35	91.35

(a) $\quad$ 4% · Price = Tax

$\qquad 0.04(\$459.00) =$

$\qquad \$18.36 = T$

$\$459.00 + \$18.36$ = Total price

$\qquad \$477.36 = TP$

(b) $\quad$ 8% · Price = Tax

$\qquad 0.08(\$66) =$

$\qquad \$5.28 = T$

$\$66.00 + \$5.28$ = Total price

$\qquad \$71.28 = TP$

(c) $\quad$ 6% · Price = Tax

$\qquad 0.06P = \$1.95$

$\qquad P = \$32.50$

$\$32.50 + \$1.95$ = Total price

$\qquad \$34.45 = TP$

(d) 5% · Price = Tax

$\qquad 0.05P = \$14.20$

$\qquad P = \$284.00$

$\$284.00 + \$14.20$ = Total price

$\qquad 298.20 = TP$

(e) 6% · Price = Tax

$\qquad 0.06P = \$3.48$

$\qquad P = \$58$

$\$58.00 + \$3.48$ = Total price

$\qquad \$61.48 = TP$

(f) $\quad P + 7\%P$ = Total price

$\qquad 1.07P = \$25.68$

$\qquad P = \$24$

$\$25.68 - \$24.00$ = Tax

$\qquad \$1.68 = T$

1. (Continued)

(g) $P + 8\%P$ = Total price

$1.08P = \$38.88$

$P = \$36$

$\$38.88 - \$36.00$ = Tax

$\$2.88 = T$

(h) $P + 5\%P$ = Total price

$1.05P = \$91.35$

$P = \$87$

$\$91.35 - \$87.00$ = Tax

$\$4.35 = T$

3. $6\% \cdot$ Price = Tax

$0.06(\$49.50) =$

$\$2.97 = T$

$\$49.50 + \$2.97$ = Total price

$\$52.47 = TP$

5. $9\% \cdot$ Price = Tax

$0.09(\$1,890) =$

$\$170.10 = T$

$\$1,890 + \$170.10$ = Total price

$\$2,060.10 = TP$

7. (a) $6\% \cdot$ Price = Tax

$0.06P = \$5.34$

$P = \$89$

(b) $\$89.00 + \$5.34$ = Total price

$\$94.34 = TP$

9. (a) $15\% \cdot$ Price = Tax

$0.15P = \$12.00$

$P = \$80$

(b) $\$80.00 + \$12.00$ = Total price

$\$92.00 = TP$

11. (a) Price + 6% Price = Total price

$1.06P = \$200.87$

$P = \$189.50$

(b) $\$200.87 - \$189.50$ = Tax

$\$11.37 = T$

13. (a) Price + 5% Price = Total price

$1.05P = \$93.45$

$P = \$89$

(b) $\$93.45 - \$89.00$ = Tax

$\$4.45 = T$

15. (a) Price + 6.5% Price = Total price

$1.065P = \$265.72$

$P = \$249.50$

(b) $\$265.72 - \$249.50$ = Tax

$\$16.22 = T$

17. (a) Price + 8% Price = Total price
          $1.08P = \$366.12$
            $P = \$339.00$

    (b) $\$366.12 - \$339.00 = Tax$
             $\$27.12 = T$

19. (a) Price + 5% Price + 15% Price = Total price
          $1.20P = \$15.60$
            $P = \$13$

    (b) 5% · Price = Sales tax
        $0.05(\$13) =$
           $\$0.65 = ST$

    (c) 15% · Price = Excise tax
          $0.15(\$13) =$
            $\$1.95 = ET$

21. (a) Price + 7% Price + 13% Price = Total price
          $1.20P = \$550.80$
            $P = \$459.00$

    (b) 7% · Price = Sales tax
        $0.07(\$459) =$
          $\$32.13 = ST$

    (c) 13% · Price = Import tax
          $0.13(\$459) =$
           $\$59.67 = IT$

23. (a) Price + 6% Price + 12% Price = Total price
          $1.18(\$25,000) =$
            $\$29,500 = TP$

    (b) 6% · Price = Sales tax
        $0.06(\$25,000) =$
          $\$1,500 = ST$

    (c) 12% · Price = Import tax
         $0.12(\$25,000) =$
          $\$3,000 = IT$

Section 2

1. (a) 1.46%
   (b) $1.46 per C
   (c) $14.58 per M
   (d) 15 mills

3. (a) 1.93%
   (b) $1.93 per C
   (c) $19.24 per M
   (d) 20 mills

5. Rate $= \dfrac{\$24,000,000}{\$980,000,000} = 0.0244897+$

    (a) 2.5%   (b) $2.45 per C   (c) $24.49 per M   (d) 25 mills

7.

	Rate	Assessed Value	Tax
(a)	2.8%	$ 94,000	$2,632
(b)	$1.25 per C	123,400	1,542.50
(c)	$12.40 per M	93,000	1,153.20
(d)	2.1%	50,000	1,050
(e)	$2.60 per C	204,000	5,304
(f)	$18.80 per M	65,000	1,222
(g)	1.5%	163,000	2,460
(h)	$3.20 per C	74,500	2,384
(i)	$9.60 per M	95,000	912

(a) R = 2.8%   R × V = T
     V = $94,000   2.8% × $94,000 =
     T = ? $2,632 = T

(b) R = $1.25 per C   R × V = T
     V = $123,400 $1.25 × $1,234 =
       = 1,234 hundred $1,542.50 = T
     T = ?

(c) R = $12.40 per M   R × V = T
     V = $93,000 $12.40 × $93 =
       = 93 thousand $1,153.20 = T
     T = ?

(d) R = 2.1%   R × V = T
     V = ?   2.1%V = $1,050
     T = $1,050   V = $50,000

(e) R = $2.60 per C   R × V = T
     V = ? $2.60V = $5,304
     T = $5,304   V = 2,040 hundred
                                      V = $204,000

(f) R = $18.80 per M   R × V = T
     V = ? $18.80V = $1,222
     T = $1,222   V = 65 thousand
                                      V = $65,000

(g) R = ?%   R × V = T
     V = $163,000   R × $163,000 = $2,460
     T = $2,460   R = 0.015 = 1.5%

7. (Continued)

(h) R = ? per M   R × V = T
    V = $74,500   R × $745 = $2,384
      = 745 hundred   R = $3.20 per C
    T = $2,384

(i) R = ? per M   R × V = T
    V = $95,000   R × $95 = $912
      = 95 thousand   R = $9.60 per M
    T = $912

9. R = 1.8%   R × V = T
   V = $120,000   0.018 × $120,000 =
   T = ? $2,160 = T

11. R = $2.05 per C   R × V = T
    V = $265,000 $2.05 × $2,650 =
      = 2,650 hundred $5,432.50 = T
    T = ?

13. R = $1.95 per C   R × V = T
    V = ? $1.95V = $7,117.50
    T = $7,117.50   V = 3,650 hundred
                                    V = $365,000

15. R = $12.75 per M   R × V = T
    V = ? $12.75V = $1,020
    T = $1,020   V = 80 thousand
                                      V = $80,000

17. R = ?%   R × V = T
    V = $96,000   R × $96,000 = $2,016
    T = $2,016   R = 0.021 = 2.1%

19. R = ? mills   R × V = T
    V = $64,000   R × $64 = $525
      = 64 thousand   R = 9 mills
    T = $525

21. R = ? per C   R × V = T
    V = $260,000   R × $2,600 = $3,960
      = 2,600 hundred   R = $1.52 per C
    T = $3,960

23. First Year
    R = ? per C   R × V = T
    V = $95,000   R × $950 = $1,349
      = 950 hundred   R = $1.42 per C
    T = $1,349

23. (Continued)

Second Year
R = $1.50 per C ($1.42 + $0.08)   R × V = T
V = ? $1.50V = $1,434
T = $1,434   V = 956 hundred
                                               V = $95,600

Increase in value = $95,600 - $95,000 = $600

25. First Year
R = 1.3%   R × V = T
V = ?   1.3%V = $390
T = $390   V = $30,000

Second year value = $30,000 + $4,000 = $34,000

Second Year
R = ?%   R × V = T
V = $34,000   R × $34,000 = $544
T = $544   R = 0.016 = 1.6%

Section 1

		Structure Value	Contents Value	Class	Territory	Premium
1.	(a)	$295,000	$64,000	B	1	$2,158
	(b)	280,000	78,000	A	2	2,154

1. (a) $0.56 \times \$2,950 = \$1,652.00$ Structure
$0.79 \times \$\ 640 = \underline{\hphantom{00}505.60}$ Contents
$\$2,157.60$
or $\$2,158$ Total

  (b) $0.58 \times \$2,800 = \$1,624.00$ Structure
$0.68 \times \$\ 780 = \underline{\hphantom{00}530.40}$ Contents
$\$2,154.40$
or $\$2,154$ Total

		Amount of Insurance	Term	Canceled By	Premium	Refund Due
3.	(a)	$ 290,000	1 year	X	$2,262	X
	(b)	770,000	4 months	X	3,003	X
	(c)	830,000	9 months	Insured	5,503	$ 971
	(d)	1,500,000	4 months	Carrier	3,900	7,800

  (a) $0.78 \times \$2,900 = \$2,262$ Annual premium

  (b) $0.78 \times \$7,700 = \$6,006$ Annual premium
For a 4 month policy: $50\% \times \$6,006 = \$3,003$

  (c) $0.78 \times \$8,300 = \$6,474$ Annual premium
Canceled by insured 9 months:
$85\% \times \$6,474 = \$5,502.90$ or $\$5,503$ Premium

    Refund $= \$6,474 - \$5,503 = \$971$

  (d) $0.78 \times \$15,000 = \$11,700$ Annual premium
Canceled by carrier after 4 months:

$$\frac{4}{12} \times \$11,700 = \$3,900 \text{ Premium}$$

    Refund $= \$11,700 - \$3,900 = \$7,800$

5.

		Property Value	Coins. Clause	Insurance Required	Insurance Carried	Amount of Loss	Indemnity
(a)	$	500,000	80%	$400,000	$200,000	$250,000	$125,000
(b)		440,000	80	352,000	352,000	90,000	90,000
(c)		600,000	90	540,000	500,000	420,000	388,889
(d)		900,000	80	720,000	600,000	680,000	566,667
(e)		1,000,000	80	800,000	700,000	950,000	700,000

(a) Insurance required = 80% × $500,000 = $400,000

$$\frac{\text{Insurance carried}}{\text{Insurance required}} = \frac{\$200,000}{\$400,000} = \frac{1}{2}$$

$\frac{1}{2}$ × $250,000 = $125,000 Indemnity

(b) Insurance required = 80% × $440,000 = $352,000
Since insurance carried equals insurance required, $90,000 is fully covered.

(c) Insurance required = 90% × $600,000 = $540,000

$$\frac{\text{Insurance carried}}{\text{Insurance required}} = \frac{\$500,000}{\$540,000} = \frac{25}{27}$$

$\frac{25}{27}$ × $420,000 = $388,889 Indemnity

(d) Insurance required = 80% × $900,000 = $720,000

$$\frac{\text{Insurance carried}}{\text{Insurance required}} = \frac{\$600,000}{\$720,000} = \frac{5}{6}$$

$\frac{5}{6}$ × $680,000 = $566,667, Indemnity.

(e) Insurance required = 80% × $1,000,000 = $800,000

$$\frac{\text{Insurance carried}}{\text{Insurance required}} = \frac{\$700,000}{\$800,000} = \frac{7}{8}$$

$\frac{7}{8}$ × $950,000 = $831,250, which exceeds the amount of

insurance carried ($700,000) so compensation will be limited to $700,000.

		Company	Amount of Policy	Ratio of Coverage	Amount of Loss	Compensation
7.	(a)	N	$440,000	22/53	$320,000	$132,830
		O	620,000	31/53		187,170
	(b)	I	$750,000	15/28	$680,000	$364,286
		N	400,000	2/7		194,286
		S	250,000	5/28		121,428

(a) Total coverage = $440,000 + $620,000 = $1,060,000

N pays $\dfrac{\$440,000}{\$1,060,000} = \dfrac{22}{53}$; $\dfrac{22}{53} \times \$320,000 = \$132,830$

O pays $\dfrac{\$620,000}{\$1,060,000} = \dfrac{31}{53}$; $\dfrac{31}{53} \times \$320,000 = \underline{\$187,170}$

$\$320,000$

(b) Total coverage = $750,000 + $400,000 + $250,000
= $1,400,000

I pays $\dfrac{\$750,000}{\$1,400,000} = \dfrac{15}{28}$; $\dfrac{15}{28} \times \$680,000 = \$364,286$

N pays $\dfrac{\$400,000}{\$1,400,000} = \dfrac{2}{7}$; $\dfrac{2}{7} \times \$680,000 = \$194,286$

S pays $\dfrac{\$250,000}{\$1,400,000} = \dfrac{5}{28}$; $\dfrac{5}{28} \times \$680,000 = \underline{\$121,428}$

$\$680,000$

		Property Value	Coins. Clause	Ins. Req.	Ins. Carried	Fire Loss
9.	(a)	$600,000	80%	$480,000	$320,000	$300,000
	(b)	800,000	70	560,000	480,000	770,000

	Total Indemnity	Policy Co. Value	Co's Ratio	Co's Payment
(a)	$200,000	I $200,000	5/8	$125,000
		J 120,000	3/8	75,000
(b)	660,000	K 180,000	3/8	247,500
		L 300,000	5/8	412,500

9. (Continued)

(a) Insurance required = 80% × $600,000 = $480,000
     Insurance carried = $200,000 + $120,000 = $320,000

$$\frac{\text{Insurance carried}}{\text{Insurance required}} = \frac{\$320,000}{\$480,000} = \frac{2}{3}$$

$\frac{2}{3}$ × $300,000 = $200,000 Total indemnity

I pays $\frac{\$200,000}{\$320,000} = \frac{5}{8}$; $\frac{5}{8}$ × $200,000 = $125,000

J pays $\frac{\$120,000}{\$320,000} = \frac{3}{8}$; $\frac{3}{8}$ × $200,000 = $ 75,000
                                                            $200,000

(b) Insurance required = 70% × $800,000 = $560,000
     Insurance carried = $180,000 + $300,000 = $480,000

$$\frac{\text{Insurance carried}}{\text{Insurance required}} = \frac{\$480,000}{\$560,000} = \frac{6}{7}$$

$\frac{6}{7}$ × $770,000 = $660,000 Total indemnity

K pays $\frac{\$180,000}{\$480,000} = \frac{3}{8}$; $\frac{3}{8}$ × $660,000 = $247,500

L pays $\frac{\$300,000}{\$480,000} = \frac{5}{8}$; $\frac{5}{8}$ × $660,000 = $412,500
                                                            $660,000

11. R = $0.60 per $100   P = R · V
    V = $650,000   = $0.60 × 6,500
      = 6,500 hundreds   P = $3,900
    P = ? (3 months)
                                40% × $3,900 = $1,560 (3 months)

13. R = $0.86 per $100   P = R · V
    V = $630,000   = $0.86 × 6,300
      = 6,300 hundreds   P = $5,418
    P = ?

    (a) Canceled by the insured:
        60% × $5,418 = $3,251 Premium
        $5,418 - $3,251 = $2,167 Refund

13. (Continued)

    (b) Canceled by the insurer:

    $$\frac{5}{12} \times \$5,418 = \$2,258 \text{ Premium}$$

    $5,418 - $2,258 = $3,160 Refund

15. Insurance required = 80% × $680,000 = $544,000
    Insurance carried = $500,000 (which exceeds insurance
    required)

    (a) $73,529
    (b) $459,559
    (c) $500,000

    (a) $\frac{Insurance\ Carried}{Insurance\ Required} \times loss = Indemnity$

    $$\frac{\$500,000}{\$544,000} \times \$80,000 = \$73,529$$

    (b) $\frac{Insurance\ Carried}{Insurance\ Required} \times loss = Indemnity$

    $$\frac{\$500,000}{\$544,000} \times \$500,000 = \$459,559$$

    (c) $\frac{Insurance\ Carried}{Insurance\ Required} \times loss = Indemnity$

    $$\frac{\$500,000}{\$544,000} \times \$600,000 = \$551,471$$ (However, insurance
    company's payment is
    limited to the face value
    of the policy, or $500,000)

17. Insurance required = 90% × $5,000,000 = $4,500,000

    $$\frac{Insurance\ carried}{Insurance\ required} = \frac{\$4,000,000}{\$4,500,000} = \frac{8}{9}$$

    (a) Indemnity = $\frac{8}{9} \times \$36,000 = \$32,000$

    (b) Indemnity = $\frac{8}{9} \times \$2,700,000 = \$2,400,000$

17. (Continued)

    (c) Indemnity = $4,000,000 (Maximum = face value)

19. Insurance required = 90% × $300,000 = $270,000

    $$\frac{\text{Insurance carried}}{\text{Insurance required}} = \frac{\$210,000}{\$270,000} = \frac{7}{9}$$

    (a) Indemnity = $\frac{7}{9}$ × $81,000 = $63,000

    (b) Indemnity = $\frac{7}{9}$ × $189,000 = $147,000

    (c) Indemnity = $210,000 (Maximum = face value)

21. Total insurance = $240,000 + $180,000 + $580,000 = $1,000,000

    A pays $\frac{\$240,000}{\$1,000,000} = \frac{6}{25}$ or 24%

    B pays $\frac{\$180,000}{\$1,000,000} = \frac{9}{50}$ or 18%

    C pays $\frac{\$580,000}{\$1,000,000} = \frac{29}{50}$ or 58%

    (a) A = 24% × $50,000 = $12,000
         B = 18% × $50,000 = $ 9,000
         C = 58% × $50,000 = $29,000
                             $50,000

    (b) A = 24% × $900,000 = $216,000
         B = 18% × $900,000 = $162,000
         C = 58% × $900,000 = $522,000
                              $900,000

    (c) A = 24% × $1,000,000 = $ 240,000
         B = 18% × $1,000,000 = $ 180,000
         C = 58% × $1,000,000 = $ 580,000
                                $1,000,000

23. Insurance carried = $300,000 + $350,000 + $250,000 + $100,000
                       = $1,000,000

23. (Continued)

AA pays $\dfrac{\$300,000}{\$1,000,000} = \dfrac{3}{10}$ or 30%

BB pays $\dfrac{\$350,000}{\$1,000,000} = \dfrac{35}{100}$ or 35%

CC pays $\dfrac{\$250,000}{\$1,000,000} = \dfrac{25}{100}$ or 25%

DD pays $\dfrac{\$100,000}{\$1,000,000} = \dfrac{1}{10}$ or 10%

(a) AA = 30% × $72,000 = $21,600
     BB = 35% × $72,000 = $25,200
     CC = 25% × $72,000 = $18,000
     DD = 10% × $72,000 = $ 7,200
                          $72,000

(b) AA = 30% × $600,000 = $180,000
     BB = 35% × $600,000 = $210,000
     CC = 25% × $600,000 = $150,000
     DD = 10% × $600,000 = $ 60,000
                           $600,000

(c) AA = 30% × $1,000,000 = $ 300,000
     BB = 35% × $1,000,000 = $ 350,000
     CC = 25% × $1,000,000 = $ 250,000
     DD = 10% × $1,000,000 = $ 100,000
                             $1,000,000

25. (a) Insurance required = 80% × $2,500,000 = $2,000,000
         Insurance carried = $1,000,000 + $800,000 = $1,800,000

$\dfrac{\text{Insurance carried}}{\text{Insurance required}} = \dfrac{\$1,800,000}{\$2,000,000} = \dfrac{9}{10}$

$\dfrac{9}{10}$ × $2,200,000 = $1,980,000 (Exceeds total face value so

$1,800,000 is total indemnity paid.)

(b) Pacific American $= \dfrac{\$1,000,000}{\$1,800,000} = \dfrac{5}{9}$ × $1,800,000 = $1,000,000

Liberty $= \dfrac{\$800,000}{\$1,800,000} = \dfrac{4}{9}$ × $1,800,000 = $800,000

27. Insurance required = $80\% \times \$6,000,000 = \$4,800,000$

    Insurance carried = $\$1,200,000 + \$1,800,000 + \$1,500,000$
    = $\$4,500,000$

$$\frac{\text{Insurance carried}}{\text{Insurance required}} = \frac{\$4,500,000}{\$4,800,000} = \frac{15}{16}$$

(a) $\dfrac{15}{16} \times \$4,500,000 = \$4,218,750$

(b) Delta pays $= \dfrac{\$1,200,000}{\$4,500,000} = \dfrac{4}{15} \times \$4,218,750 = \$1,125,000$

   Eagle pays $= \dfrac{\$1,800,000}{\$4,500,000} = \dfrac{6}{15} \times \$4,218,750 = \$1,687,500$

   Franklin pays $= \dfrac{\$1,500,000}{\$4,500,000} = \dfrac{1}{3} \times \$4,218,750 = \$1,406,250$

Section 2

1. (a)

	Territory 3
$115.00	25/25 Bodily injury
103.00	$10,000 Property damage
64.00	$1,000 Medical pay
$282.00	Base premium
× 1.20	Driver classification
$338.40	Annual premium

   (b)

	Territory 1
$ 88.00	50/50 Bodily injury
86.00	$25,000 Property damage
67.00	$5,000 Medical pay
$241.00	Base premium
× 1.30	Driver classification
$313.30	Annual premium

   (c)

	Territory 1
$ 94.00	100/200 Bodily injury
87.00	$50,000 Property damage
70.00	$10,000 Medical pay
$251.00	Base premium
× 1.65	Driver classification
$414.15	Annual premium

3. (a) $ 64.00   Comprehensive (Model M, Age 3, Territory 1)
       125.00 $250-deductible collision
      $189.00   Base premium
      × 1.40   Driver classification
      $264.60   Annual premium

   (b) $ 69.00   Comprehensive (Model J, Age 1, Territory 2)
       108.00 $500-deductible collision
      $177.00   Base premium
      × 1.90   Driver classification
      $336.30   Annual premium

   (c) $ 58.00   Comprehensive (Model D, Age 2, Territory 3)
        85.00 $500-deductible collision
      $143.00   Base premium
      × 1.40   Driver classification
      $200.20   Annual premium

5.   Territory 1
   $ 86.00   25/50 Bodily injury
     85.00 $10,000 Property damage
     62.00 $1,000 Medical pay
   $233.00   Base premium
   × 1.70   Driver classification (Male, 20, not principal
              operator, D.T., 6 miles)
   $396.10   Annual premium

7. $ 77.00   Comprehensive (Model N, Age 1, Territory 1)
     140.00 $250-deductible collision
   $217.00   Base premium
   × 1.50   Driver classification (Female, 27, uses in
              business)
   $325.50   Annual premium

9.   Territory 2
   $104.00   100/100 Bodily injury
    101.00 $50,000 Property damage
     72.00 $10,000 Medical pay
     56.00   Comprehensive (Model B, Age 2)
     86.00 $250-deductible collision
   $419.00   Base premium
   × 1.00   Driver classification (No young operator, 2½ miles)
   $419.00   Annual premium

11.
$ 94.00	25/25 Bodily injury
97.00	$10,000 Property damage
66.00	$2,500 Medical pay
57.00	Comprehensive (Model K, Age 4)
91.00	$500-deductible collision
$405.00	Base premium
× 1.40	Driver classification (No young operator, 11 miles)
$567.00	Annual premium

Territory 2 (heading above the list)

13.

	Damages/ Court Award	Insurance Company Pays	Insured Pays
Person A	$ 30,000	$30,000	$ 0
Person B	65,000	50,000	15,000
Car	25,000	10,000	15,000
	$120,000	(a) $90,000	(b) $30,000

15. (a) $16,500 should be covered by Bruce's comprehensive insurance (unless the policy had a deductible for comprehensive coverage).

(b) Replacement Cost - Indemnity = Out-of-Pocket.
$20,000 - $16,500   = $3,500

17.

	Damages/ Court Award	Insurance Company Pays	Insured Pays
Driver	$ 48,000	$25,000	$23,000
Passenger	60,000	25,000	35,000
Car	24,000	5,000	19,000
Van	7,000	6,500	500
	$139,000	(a) $61,500	(b) $77,500

19.

	Damages/ Court Award	Insurance Company Pays	Insured Pays
Widow	$400,000	$100,000	$300,000
Gemini	110,000	100,000	10,000
Edmund's doctor bills	3,000	1,000	2,000
Car	15,000	15,000	0
Truck	2,000	1,750	250
	$530,000	(a) $217,750	(b) $312,250

Section 3

1.

	Applicant and Age	Type of Policy	Face Value	Annual Premium
(a)	Male, 22	10-year term	$ 15,000	$ 104.25
(b)	Male, 35	Whole life	50,000	1,199.50
(c)	Female, 30	20-pmt. life	20,000	561.20
(d)	Female, 35	Variable universal	100,000	2,605.00

(a) $ 6.95   (b) $ 23.99   (c) $ 28.06   (d) $ 26.05
    ×   15   ×   50   ×   20   ×   100
    $104.25 $1,199.50 $561.20 $2,605.00

3.

	Yrs. in Force	Type of Policy	Face Value	Nonforfeiture Option	Nonforfeiture Value
(a)	20	Whole life	$100,000	Cash value	$25,100
(b)	5	20-pay. life	50,000	Paid-up ins.	$10,900
(c)	10	Whole life	125,000	Extend. term	18 yrs. 91 days
(d)	5	Var. universal	150,000	Cash value	$14,250

(a) $ 251   (b) $ 218   (c) 18 years and 91 days
    ×   100   ×   50
    $25,100 $10,900

(d) $ 95
    ×   150
    $14,250

5.

	Beneficiary			Settlement	Monthly Annuity	
	Sex	Age	Face Value	Option Chosen	Years	Amount
(a)	M	60	$ 55,000	Fixed no. of years	16	$353.65
(b)	F	50	75,000	Fixed amt. per mo.	20	425.00
(c)	F	65	100,000	Life annuity	--	656.00
(d)	M	45	150,000	Guaranteed annuity	20	720.00

(a) $ 6.43   (b) $\frac{425}{75} = 5.66\overline{6}$ so 20 years
    ×   55
    $353.65

(c) $ 6.56   (d) $ 4.80
    ×   100   ×   150
    $656.00 $720.00

7. (a) <u>20-payment life:</u>
$ 25.59 (age 20)
× 100
$2,559.00 Annual premium

(b) <u>Whole life:</u>
$ 15.46 (age 20)
× 100
$1,546.00 Annual premium

(c) $2,559.00 Annual premium
× 20 years (paid up)
$ 51,180 Total premium

(d) $1,546.00 Annual premium
× 30 years
$ 46,380 Total premium

9. (a) <u>10-year term:</u>
$ 7.10 (set back to 25)
× 50
$355.00 Annual premium

(b) $ 355 Annual premium
× 10 years
$3,550 Total premium

(c) <u>Variable Universal:</u>
$ 22.17 (set back to 25)
× 50
$ 1,108.50 Annual premium
× 10 years
$11,085.00 Total premium

(d) $8,610 Whole life
-3,550 10-year term
$5,060 Savings

(e) Term = 0; Variable Universal = $50,000

11. (a) <u>At age 30:</u>
$ 19.73 Whole life
× 75
$1,479.75 Annual premium

<u>At age 25:</u>
$ 17.22 Whole life
× 75
$1,291.50 Annual premium

$1,479.75 Annual premium at age 30
-1,291.50 Annual premium at age 20
$ 188.25 Annual savings

(b) $ 1,479.75 Annual premium
× 35 years
$51,791.25 Total premiums at age 30

(c) $ 1,291.50 Annual premium
× 40 years
$51,660.00 Total premiums at age 20

(d) If he takes out the policy at age 25, he gets 5 more years of insurance protection, and the policy builds cash value during those extra 5 years.

13. (a) <u>Annual premium</u>:

    $28.06
    ×   100
    $2,806

    (b) <u>Cash value</u>:

    $ 187
    ×   100
    $18,700

    (c) <u>Paid-up</u>:

    $ 507
    ×   100
    $50,700

    (d) 28 years and 186 days for extended term

15. $ 319
    ×   120
    $38,280 Cash value

17. (a) <u>Variable universal</u>:

    $ 1,030
    ×   100
    $103,000 Cash value

    (b) <u>Whole life</u>:

    $ 251
    ×   100
    $25,100 Cash value

19. (a) <u>Fixed # of years = 14</u>:

    $ 7.71
    ×   155
    $1,195.05 per month

    (b) <u>Fixed amount</u>:

    $$\frac{\$1,000}{155} = 6.452$$

    Closest table value = $6.43;
    so $1,000 per month payments
    would last approximately 16
    years

21. (a) <u>Life annuity</u>:

    $ 5.30
    ×   125
    $662.50/month for life

    (b) <u>Life annuity guaranteed for
    20 years</u>:

    $ 5.00
    ×   125
    $625.00 per month

23. (a) <u>Fixed # of years = 10</u>:

    $ 9.60
    ×   50
    $ 480 per month
    ×   120 months
    $57,600 Total annuity

    (b) <u>Life annuity with 10-years
    certain</u>:

    $ 5.28
    ×   50
    $ 264
    ×   156 months
    $41,184

    (c) She gained $16,416 (and gained $7,600 over face value).

25. (a) <u>Life annuity:</u>
    $ 5.30
    ×   85
    $450.50 per month

   (b) $ 450.50
     ×   144 months
     $64,872.00 Total received

   (c) <u>Life annuity guaranteed 20 years:</u>
    $ 5.00
    ×   85
    $425.00 per month

        $ 425
        ×   144 months
        $61,200 Total received

   (d) Jennifer gained $3,672 ($64,872 − $61,200), but secondary beneficiary lost $40,800 ($425 × 96 months).

Section 1

1.

```
#360    Balance
        Brought
        Forward  $3,645.26
Nov. 2 20 xx
To  Gold Supply
    Co.
For Office supplies,
    $176, plus sales
    tax, $7.92
        Amount
        Check     $  183.92
        Deposit    --------
        Balance    $3,461.34
```

```
#361    Balance
        Brought
        Forward  $3,461.34
Nov. 3 20 xx
To  Baltimeier
    Computer Co.
For Computer,
    $940, plus
    sales tax, $42.30
        Amount
        Check     $  982.30
        Deposit    --------
        Balance    $2,479.04
```

```
#362    Balance
        Brought
        Forward  $2,479.04
Nov. 3 20 xx
To:  Garman, Inc.


For: Merchandise


        Amount
        Check     $2,190.00
        Deposit   $  374.46
        Balance   $  663.50
```

```
#363    Balance
        Brought
        Forward   $663.50
Nov. 4 20 xx
To:   U.S. Postal
      Service
For:  Postage
      stamps


        Amount
        Check     $   64.00
        Deposit   $  500.00
        Balance   $1,099.50
```

3.

Bank Reconciliation			
Bank Balance	$1,205.75	Checkbook Balance	$742.12
Less: Outstanding checks			
$ 93.20			
122.48			
250.39		Less:	
12.56	-478.63	Service charge	-15.00
Adjusted Balance	$727.12	Adjusted Balance	$727.12

5.

Bank Reconciliation			
Bank Balance	$1,814.41	Checkbook Balance	$1,077.97
Add: Outstand. deposit	+ 324.37		
	$2,138.78		
Less: Outstand. checks		Less:	
$176.00		Service charge $10.45	
440.60		Returned check 50.00	- 60.45
504.66	-1,121.26		
Adjusted Balance	$1,017.52	Adjusted Balance	$1,017.52

7.

Bank Reconciliation			
Bank Balance	$ 845.32	Checkbook Balance	$1,348.29
Add: Outstand. deposit	+ 800.00	Add: Interest	+ 8.48
	$1,645.32		$1,356.77
		Less:	
		Correction to	
		deposit charge $ 2.00	
		Service charge 6.55	
		Returned check 165.65	
Less: Outstand. checks	- 477.75	Printing charge 15.00	- 189.20
Adjusted Balance	$1,167.57	Adjusted Balance	$1,167.57

9.

Bank Reconciliation			
Bank Balance	$2,675.71	Checkbook Balance	$1,352.18
Add: Outstand. deposit	+ 836.24		
	$3,511.95		
Less: Outstand. checks			
#224 $ 376.28			
#229 54.62		Less:	
#232 282.44		Returned check $400.00	
#233 1,856.43	-2,569.77	Service charge 10.00	- 410.00
Adjusted Balance	$ 942.18	Adjusted Balance	$ 942.18

11.

Bank Reconciliation			
Bank Balance	$1,525.34	Checkbook Balance	$1,079.27
Add: Outstand. deposit	+ 234.03		
	$1,759.37		
Less: Outstand. checks			
#231 $ 78.33		Less:	
#237 262.13		Bank charges $18.45	
#239 4.34		7.40	- 25.85
#242 361.15	- 705.95		
Adjusted Balance	$1,053.42	Adjusted Balance	$1,053.42

Section 2

1.

Kitchens, Inc. Daily Cash Report		
Date: Register: No. 9 Clerk: Tom Bruno		
Pennies	52	$ 0.52
Nickels	68	3.40
Dimes	30	3.00
Quarters	74	18.50
Halves	2	1.00
Ones	34	34.00
Fives	12	60.00
Tens	15	150.00
Twenties	10	200.00
Other currency ($100)	1	100.00
Checks (listed separately):		
	$50.25	
	36.36	
	15.98	
	48.00	
	71.45	+222.04
Total cash		$792.46
Less: change fund		-100.00
Net cash receipts		$692.46
Cash over (subtract)		- 2.98
Cash short (add)		-------
Cash register total		$689.48

$2.98 over

3.

```
┌─────────────────────────────────────────────────┐
│                  Hair, Inc.                      │
│              Daily Cash Report                   │
├─────────────────────────────────────────────────┤
│ Date:                                            │
│ Register:                                        │
│ Clerk:                                           │
├─────────────────────────────────────────────────┤
│ Pennies                          22  $    0.22   │
│ Nickels                          74       3.70   │
│ Dimes                            53       5.30   │
│ Quarters                         15       3.75   │
│ Halves                            1       0.50   │
│ Ones                             36      36.00   │
│ Fives                            18      90.00   │
│ Tens                             12     120.00   │
│ Twenties                          5     100.00   │
│ Other currency ($50)              1      50.00   │
│ Checks (listed                                   │
│ separately):           $25                       │
│                         30                       │
│                         42                       │
│                         25.60                    │
│                         35           +157.60     │
│                                                  │
│ Total Cash                          $567.07      │
│    Less:  change fund               - 75.00      │
│    Net cash receipts                $492.07      │
│    Cash over (subtract)             -------      │
│    Cash short (add)                 +   4.15     │
│    Cash register total              $496.22      │
└─────────────────────────────────────────────────┘
```

Cash short $4.15

5.

USA Chips Co. Over and Short Summary				
Date	Total Sales	Net Cash Receipts	Cash Over	Cash Short
June 3	$1,045.68	$1,045.70	$0.02	--
June 4	1,172.49	1,172.49	--	--
June 5	1,099.46	1,100.00	0.54	--
June 6	1,153.33	1,153.25	--	$0.08
June 7	1,155.21	1,155.16	--	0.05
Totals	$5,626.17	$5,626.60	$0.56	$0.13

Total cash receipts $5,626.60
Total cash short (add)   +
      0.13

Total cash over (subtract)   −   0.56
Total cash register readings $5,626.17

Cash over $0.56
Cash short   −0.13
Cash over $0.43 for the week

CHAPTER 8 - WAGES AND PAYROLLS
PROBLEM SOLUTIONS

Sections 1 and 2

		ANNUAL SALARY	MONTHLY	SEMIMONTHLY	WEEKLY	BIWEEKLY
1.	(a)	$ 48,000	$ 4,000	$2,000	$ 923.08	$1,846.15
	(b)	72,000	6,000	3,000	1,384.62	2,769.23
	(c)	24,000	2,000	1,000	461.54	923.08
2.	(a)	$150,000	$12,500	$6,250	$2,884.62	$5,769.23
	(b)	33,600	2,800	1,400	646.15	1,292.31
	(c)	96,000	8,000	4,000	1,846.15	3,692.31

1. (a) $\dfrac{\$48,000}{12} = \$4,000/\text{monthly}$

$\dfrac{\$48,000}{24} = \$2,000/\text{semimonthly}$

$\dfrac{\$48,000}{52} = \$923.08/\text{weekly}$

$\dfrac{\$48,000}{26} = \$1,846.15/\text{biweekly}$

(b) $\$6,000 \times 12 = \$72,000/\text{annual}$

$\dfrac{\$72,000}{24} = \$3,000/\text{semiannual}$

$\dfrac{\$72,000}{52} = \$1,384.62/\text{weekly}$

$\dfrac{\$72,000}{26} = \$2,769.23/\text{biweekly}$

1. (Continued)

    (c) $1,000 × 24 = $24,000/annual

$$\frac{\$24,000}{12} = \$2,000/\text{monthly}$$

$$\frac{\$24,000}{52} = \$461.54/\text{weekly}$$

$$\frac{\$24,000}{26} = \$923.08/\text{biweekly}$$

		SALARY	QUOTA	RATE	NET SALES	COMMISSION	GROSS WAGE
3.	(a)	X	X	6 %	$19,300	$1,158	X
	(b)	X	X	5½	27,600	1,518	X
	(c)	X	X	4	61,000	2,440	X
	(d)	$2,000	X	5	44,000	2,200	$4,200
	(e)	1,100	X	7	60,000	4,200	5,300
	(f)	X	$5,000	4½	80,000	3,375	X
	(g)	1,750	2,000	8	12,500	840	2,590
	(h)	800	1,000	6	52,000	3,060	3,860
	(i)	2,200	1,400	5	13,800	620	2,820

3. (a)   % × S = C
    6% × $19,300 =
    $1,158 = C

(b)   % × S = C
    5.5% × S = $1,518
    S = $27,600

(c)   % × S = C
    X($61,000) = $2,440
    X = 4%

(d)   % × S = C
    5% × $44,400 =
    $2,200 = C

Salary + Commission = Gross Wage
$2,000 + $2,200 = GW
$4,200 = GW

3. (Continued)

   (e) Salary + Commission = Gross wage
                $1,100 + C = $5,300
                         C = $4,200

        % × S = C
       7% × S = $4,200
            S = $60,000

   (f) Net sales - Quota = Net sales subject to commission
        $80,000 - $5,000 = $75,000

               % × S = C
       4.5% × $75,000 =
            $3,375 = C

   (g) Net sales - Quota = Net sales subject to commission
        $12,500 - $2,000 = $10,500

               % × S = C   Salary + Commission = Gross
       8% × $10,500 =   Wage
            $840 = C $1,750 + $840 = GW
                                             $2,590 = GW

   (h) Salary + Commission = Gross wage
                 $800 + C = $3,860
                        C = $3,060

        % × S = C
       6% × S = $3,060
            S = $51,000

       $51,000 Net sales subjected to commission
        +1,000 Quota
       $52,000 Net sales

   (i) $13,800 Net sales
        - 1,400 Quota
       $12,400 Net sales subject to Commission

               % × S = C $2,820 Gross wage
       5% × $12,400 =   -   620 Commission
            $620 = C $2,200 Salary

5. $7,500 Sales
    -   350 Sales returns
    $7,150 Net sales

                        % × S = C
                9% × $7,150 =
                        $643.50 = C

7.   % × S = C
    6% × S = $10,500
          S = $175,000

9. (a) $13,600 Net sales
        - 2,000 × 3.5% = $ 70
        $11,600
        - 4,000 × 5%   =   200
        $ 7,600 × 6%   =   456
                              $726

    (b) $726 Commission
        -350 Drawing account
        $376 Amount due

11. $6,300 Gross wage
    -1,200 Salary
    $5,100 Commission

                        % × S = C
                5% × S = $5,100
                      S = $102,000

13. $7,800 Sales
    -   400 Sales returns
    $7,400 Net sales

    $7,400 Net sales
    -1,000 Quota
    $6,400 Net sales subject
                to commission

            % × S = C
    4% × $6,400 =
            $256 = C

    $500 Salary
    +256 Commission
    $756 Gross wage

15. Marshall's net sales $6,200
    Quota   -   500
    Net sales subject to commission $5,700

            % × S = C
    6% × $5,700 =
            $342 = C

    Department sales $10,100
    Sales returns   -   50
    Net sales subject to override $10,050

            % × S = C
    1.5% × $10,050 =
            $150.75 = C

    Salary $ 600
    Commission + 342
    Override   + 150.75
                $1,092.75

17.	OZARK WHOLESALE COMPANY Payroll for Week Ending June 11, 20XX								
NAME	NET SALES							COMM. RATE	GROSS COMM.
	M	T	W	TH	F	S	TOTAL		
Dempsey, J.	$3,960	$3,750	$2,800	$1,950	$3,040	$4,000	$19,500	8%	$1,560.00
Gold, K.	2,266	4,830	1,501	1,442	1,455	3,423	14,917	10	1,491.70
Keller, E.	2,322	3,880	2,404	2,425	2,421	4,500	17,952	9	1,615.68
Miller, M.	2,329	4,284	2,920	2,750	2,360	4,100	18,743	8	1,499.44
Wright, C.	2,340	4,650	2,488	1,492	1,502	2,613	15,085	7	1,055.95
								Total	$7,222.77

19.

ROZELLA COMPUTER CO.
Payroll for Week Ending October 4, 20XX

| NAME | SALES | | | QUOTA | COMM. SALES | COMM. RATE | GROSS COMM. | SALARY | GROSS WAGES |
	GROSS	R&A	NET						
Bellis, E.	$63,400	$400	$63,000	--	$63,000	8%	$ 5,040	$ 800	$ 5,840
Chambers, D.	58,200	700	7,500	$ 8,000	49,500	6	2,970	--	2,970
Duggan, C.	46,900	500	46,400	--	46,400	5	2,320	1,000	3,320
Meyer, C.	65,000	--	65,000	10,000	55,000	6	3,300	1,200	4,500
Quader, E.	63,320	820	62,500	5,000	7,500	8	4,600	--	4,600
Total							$18,320	$3,000	$21,230

21. 220 lbs. × $0.65 = $ 143
   175 lbs. × 0.52 = 91
   460 lbs. × 0.48 = 220.80
   200 lbs. × 6.30 = 1,260
   Gross proceeds $1,714.80
  Freight $ 98.40
  Insurance 100.00
  Commission 102.89
        − 301.29
  Net proceeds $1,413.51

23.   5 × $70 = $ 350
   10 × 36 =   360
    3 × 61 =   183
    2 × 57 =   114
    4 × 45 =   180
   Prime cost $1,187.00
  Shipping $96
  Insurance 55
  Commission 59.35
        + 210.35
  Gross cost $1,397.35

Section 3

1. CRISTO CLEANING SERVICE Payroll for Week Ending May 5, 20XX									
Name	M	T	W	TH	F	S	Total Hours	Rate Per Hour	Gross Wages
Becker, P.	8	8	7	7	6	0	36	$16.50	$ 594.00
Doyle, D.	8	7	8	9	8	0	40	20.00	800.00
Margo, D.	5	8	7	8	8	3	39	10.50	409.50
Murray, A.	8	7	7	7	7	2	38	13.00	494.00
Richards, B.	6	6	6	6	6	6	36	12.50	450.00
Stark, R.	8	9	9	8	6	0	40	15.00	600.00
Wu, Y.	9	8	8	7	5	0	37	16.00	592.00
								Total	$3,939.50

3.

SAUNDERS, INC.

Payroll for Week Ending December 6, 20XX

Name	M	T	W	TH	F	S	Total Hours	Reg. Hours	Hourly Rate	Over-time Hours	Over-time Rate	Reg. Wages	Over-time Wages	Total Gross Wages
Agnew	8	8	8	8	8	0	40	40	$17.50	—	$26.25	$ 700	—	$ 700.00
Barski	8	7	9	8	7	3	42	40	16.50	2	24.75	660	$ 49.50	709.50
Clancy	6	8	7	8	8	0	37	37	16.00	—	24.00	592	—	592.00
Ibar	7	8	8	9	8	5	45	40	17.20	5	25.80	688	129.00	817.00
Nozek	8	8	6	9	8	4	43	40	17.00	3	25.50	680	76.50	756.50
Totals												$3,320	$255.00	$3,575.00

5.

Name		M	T	W	TH	F	S	S	Total Hours	Rate Per Hour	Base Wages	Total Gross Wages
Coffey	RT	6	8	7	8	7	0	0	36	$ 8.00	$228.00	
	1½	0	0	0	1	0	0	0	1	12.00	12.00	
	DT	0	0	0	0	0	0	4	4	16.00	64.00	$ 364.00
Hanson	RT	7	8	8	8	7	2	0	40	12.50	500.00	
	1½	0	0	2	0	0	6	0	8	18.75	150.00	
	DT	0	0	0	0	0	0	0	0	25.00	--	650.00
Yoder	RT	8	5	8	8	8	3	0	40	15.00	600.00	
	1½	1	0	0	1	0	1	0	3	22.50	67.50	
	DT	0	0	0	0	0	0	3	3	30.00	90.00	757.50
											Total	$1,771.50

		POSITION	REGULAR HOURS PER WEEK	REGULAR SALARY	HOURS WORKED	GROSS WAGES
7.	(a)	Sales manager	40	$336	45	$399.00
	(b)	Office manager	38	304	44	368.00
	(c)	Plant manager	(varies)	437	46	465.50

7. (a) $\dfrac{\$336}{40}$ = $8.40 regular hourly rate

$$
\begin{array}{rll}
40 \times \$\ 8.40 & = & \$336 \\
5 \times 12.60 & = & \underline{63} \\
& & \$399
\end{array}
$$

(b) $\dfrac{\$304}{38}$ = $8.00 regular hourly rate

$$
\begin{array}{rll}
40 \times \$\ 8.00 & = & \$320 \\
4 \times 12.00 & = & \underline{48} \\
& & \$368
\end{array}
$$

7. (Continued)

(c) $\dfrac{\$437}{46}$ = $9.50 regular hourly rate

```
40 ×  $ 9.50 = $380
 6 ×   14.25 =   85.50
               $465.50
```

		EMPLOYEE	HOURS	RATE PER HOUR	GROSS EARNINGS
9.	(a)	G	43	$14.00	$623.00
	(b)	H	45	7.70	365.75
	(c)	I	47	8.50	429.25

9. (a) (1) 40 × $14.00 = $560.00   (2) 43 × $14.00 = $602.00
        3 ×   21.00 =   63.00   3 ×   7.00 =   21.00
                 $623.00 $623.00

    (b) (1) 40 × $ 7.70 = $308.00   (2) 45 × $7.70 = $346.50
        5 ×   11.55 =   57.75   5 ×   3.85 =   19.25
                 $365.75 $365.75

    (c) (1) 40 × $ 8.50 = $340.00   (2) 47 × $8.50 = $399.50
        7 ×   12.75 =   89.25   7 ×   4.25 =   29.75
                 $429.25 $429.25

11.

KING APPLIANCES, INC.
Payroll for Week Ending February 15, 20XX

Name	M	T	W	TH	F	S	Total Hours	Rate Per Hour	Over-time Hours	Over-time Excess Rate	Regular Rate Wages	Over-time Excess Wages	Total Gross Wages
Dinh	8	8	10	8	9	5	48	$8.50	8	$4.25	$ 408.00	$ 34.00	$ 442.00
Horn	8	9	8	9	9	6	49	9.80	9	4.90	480.20	44.10	524.30
Rook	7	8	10	10	8	4	47	7.30	7	3.65	343.10	25.55	368.65
Soka	9	9	8	8	8	0	42	7.90	2	3.95	331.80	7.90	339.70
Vern	8	9	9	6	10	8	50	7.60	10	3.80	380.00	38.00	418.00
Totals											$1,943.10	$149.55	$2,092.65

1.

	M	T	W	TH	F	Total
Stone, Tom	161	140	155	158	167	781

Base production	500 × $0.52 =	$260.00
Premium production	281 × $0.70 =	196.70
Gross wage		$456.70

3.

	M	T	W	TH	F	Total
Gross production	325	320	346	350	342	1,683
Less: Spoilage	5	6	2	0	1	14
Net production	320	314	344	350	341	1,669

Net production	1,669		
Base production	−1,000	× $0.15 =	$150.00
Premium production	669	× 0.24 =	160.56
			$310.56
Docking	14	× $0.08 =	− 1.12
Gross wages			$309.44

5.	PACIFIC TOOL CO. Payroll for Week Ending April 5, 20XX							
	NET PIECES PRODUCED					Total Net Pieces	Rate Per Piece	Gross Wages Earned
Name	M	T	W	TH	F			
Abbott, A.	64	78	76	84	90	392	$1.00	$ 392.00
Breck, J.	71	77	78	79	86	391	1.03	402.73
Dunn, M.	56	58	62	65	67	308	1.01	311.08
Jones, B.	80	89	88	85	97	439	1.04	456.56
Simon, C.	75	75	82	86	89	407	1.02	415.14
Woods, V.	84	85	89	86	91	435	1.05	456.75
Yoe, T.	88	85	88	90	92	443	1.06	469.58
							Total	$2,903.84

7.

STROZIER MANUFACTURING CO.
Payroll for Week Ending January 18, 20XX

Name	NET PIECES PRODUCED						Reg. Time Prod.	Rate Per Piece	Over-time Prod.	Over-time Rate	Reg. Wages	Over-time Wages	Total Gross Wages
	M	T	W	TH	F	S							
Chang	56	58	60	62	66	50	302	$0.80	50	$1.20	$ 241.60	$ 60.00	$ 301.60
Evans	60	61	64	67	70	45	322	0.68	45	1.02	218.96	45.90	264.86
Hall	72	75	75	79	78	70	379	0.70	70	1.05	265.30	73.50	338.80
Long	65	70	76	71	75	63	357	0.74	63	1.11	264.18	69.93	334.11
Thomas	61	71	73	76	74	68	355	0.68	68	1.02	241.40	69.36	310.76
										Totals	$1,171.04	$303.69	$1,550.13

9.

SCHNELL ATHLETIC EQUIPMENT

Payroll for Week Ending August 12, 20XX

Name	DAILY PRODUCTION					Total Prod.	Net Quota	Production	Rate	Base Wages	Total Gross Wages
	M	T	W	TH	F						
Allen GP	39	48	50	51	46		200	Base:	$0.88	$176.00	
S	1	2	0	1	2			Premium:	0.95	26.60	
NP	38	46	50	50	44	228		Chargeback:	0.75	(4.50)	$ 198.10
Dill GP	36	41	49	54	50		150	Base:	0.92	138.00	
S	0	2	2	0	0			Premium:	0.99	75.24	
NP	36	39	47	54	50	226		Chargeback:	0.60	(2.40)	210.84
Edwards GP	54	58	64	60	63		250	Base:	0.70	175.00	
S	3	1	1	0	0			Premium:	0.80	35.20	
NP	51	57	63	60	63	294		Chargeback:	0.50	(2.50)	207.70
Jenks GP	48	49	53	59	66		300	Base:	0.85	230.35	
S	1	0	2	1	0			Premium:	0.93	---	
NP	47	49	51	58	66	271		Chargeback:	0.40	(1.60)	228.75
Strong GP	65	68	73	75	79		275	Base:	0.86	236.50	
S	3	1	1	2	1			Premium:	0.96	73.92	
NP	62	67	72	73	78	352		Chargeback:	0.45	(3.60)	306.82
										Total	$1,152.21

1.

LESSIN PRINTING CO.
Payroll for Week Ending February 3, 20XX

Employee (Addl. Deps.)	Allowances	Gross Wages	DEDUCTIONS					Net Wages Due
			Social Security	Medi-care	Fed. Inc. Tax	Other (Ins.)	Total Deductions	
Beasley (0)	M-2	$ 545	$ 33.79	$ 7.90	$ 27	$10	$ 78.69	$ 466.31
Kirtley (1)	M-2	860	53.32	12.47	75	14	154.79	705.21
Mason (0)	S-1	772	47.86	11.19	105	10	174.05	597.95
Musser (2)	M-3	751	46.56	10.89	50	18	125.45	625.55
Smith (3)	M-4	687	42.59	9.96	30	22	104.55	582.45
Totals		$3,615	$224.12	$52.41	$287	$74	$637.53	$2,977.47

3.

GRAYSTONE HARDWARE CO.
Payroll for Week Ending June 14, 20XX

Empl. No.	Allowances	Total Hours	Rate Per Hour	Gross Wages	DEDUCTIONS					Net Wages Due
					Social Security	Medicare	Fed. Income Tax	Other (Union Dues)	Total Deds.	
44	S-1	40	$13.50	$ 540.00	$ 33.48	$ 7.83	$ 58	$ 8.00	$107.31	$ 432.69
45	M-1	39	14.50	565.50	35.06	8.20	39	8.50	90.76	474.74
46	M-2	40	14.00	560.00	34.72	8.12	30	8.50	81.34	478.66
47	S-2	40	15.75	630.00	39.06	9.14	63	9.00	120.20	509.80
48	S-1	38	15.00	570.00	35.34	8.27	63	9.00	115.61	454.39
			Totals	$2,865.50	$177.66	$41.56	$253	$43.00	$515.22	$2,350.28

5.

SWARTS SALES CORP.
Payroll for Week Ending February 17, 20XX

Empl. No.	Allow-ances	Net Sales	Comm. Rate	Gross Wages	DEDUCTIONS					Net Wages Due
					Social Security	Medi-care	Fed. Income Tax	Other (United Way)	Total Deductions	
G-4	S-0	$7,600	8.5%	$ 646	$ 40.05	$ 9.37	$ 88	$ 64.60	$202.02	$ 443.98
G-5	M-2	7,500	7	525	32.55	7.61	25	52.50	117.66	407.34
G-6	M-1	7,800	8	624	38.69	9.05	48	62.40	158.14	465.86
G-7	M-3	8,000	8.5	680	42.16	9.86	39	68.00	159.02	520.98
G-8	S-1	7,400	7.5	555	34.41	8.05	60	55.50	157.96	397.04
			Totals	$3,030	$187.86	$43.94	$260	$303.00	$794.80	$2,235.20

7.

KING ASSOCIATES, INC.
Payroll for Week Ending June 11, 20XX

Employee No.	Allow-ances	Net Prod.	Rate Per Piece	Gross Wages	DEDUCTIONS					Net Wages Due
					Social Security	Medicare	Fed. Income Tax	Other (Ins.)	Total Deductions	
54(2)	S-2	6,200	$0.10	$ 620	$ 38.44	$ 8.99	$ 61	$20	$128.43	$ 491.57
55(1)	M-2	5,800	0.12	696	43.15	10.09	50	15	118.24	577.76
56(2)	M-3	4,400	0.15	660	40.92	9.57	36	20	106.49	553.51
57(0)	S-1	5,000	0.13	650	40.30	9.43	75	10	134.73	515.27
58(1)	S-1	4,000	0.16	640	39.68	9.28	73	15	136.96	503.04
			Totals	$3,266	$202.49	$47.36	$295	$80	$624.85	$2,641.15

9.

MCVEIGH INDUSTRIES
Payroll for Week Ending October 8, 20XX

Empl.	Allow-ances	Hours Worked	Reg. Wages	Over-time Wages	Total Gross Wages	DEDUCTIONS					Net Wages Due
						Social Security	Medi-care	Fed. Income Tax	Other (Ins.)	Total Deductions	
E	M-3	41	$ 700	$ 26.25	$ 726.25	$ 45.03	$10.53	$ 45	$10	$110.56	$ 615.69
F	M-2	45	600	112.50	712.50	44.18	10.33	53	10	117.51	594.99
G	S-0	48	580	174.00	754.00	46.75	10.93	115	10	182.68	571.32
H	S-1	43	660	74.25	734.25	45.52	10.65	95	10	161.17	573.08
Totals			$2,540	$387.00	$2,927.00	$181.48	$42.44	$308	$40	$571.92	$2,355.08

11.

DAWES EQUIPMENT, INC.
Payroll for Week Ending January 5, 20XX

Empl.	Allow-ances	Net Reg. Produc-tion	Over-time Produc-tion	Reg. Piece Rate	Reg. Wages	Over-time Wages	Total Gross Wages	DEDUCTIONS					Net Wages Due
								Social Security	Medi-care	Fed. Inc. Tax	Other (Ins.)	Total Deduc-tions	
#20	M-2	890	75	$0.60	$ 534	$ 67.50	$ 601.50	$ 37.29	$ 8.72	$ 36	$ 30.08	$112.09	$ 489.41
21	M-0	850	50	0.64	544	48.00	592.00	36.70	8.58	52	29.60	126.88	465.12
22	S-2	825	35	0.72	594	37.80	631.80	39.17	9.16	63	31.59	142.92	488.88
23	M-1	800	45	0.88	704	59.40	763.40	47.33	11.07	69	38.17	165.57	597.83
				Totals	$2,376	$212.70	$2,588.70	$160.49	$37.53	$220	$129.44	$547.46	$ 2,041.24

Section 6

Problem 1

2	Total wages and tips, plus other compensation	**2**	15,000		
3	Total income tax withheld from wages, tips, and sick pay	**3**	2,250		
4	Adjustment of withheld income tax for preceding quarters of calendar year	**4**	—		
5	Adjusted total of income tax withheld (line 3 as adjusted by line 4 — see instructions) . .	**5**	2,250		
6	Taxable social security wages . . . . **6a** 15,000 × 12.4% (.124) =	**6b**	1,860		
	Taxable social security tips . . . . **6c** — × 12.4% (.124) =	**6d**			
7	Taxable Medicare wages and tips . . . . **7a** 15,000 × 2.9% (.029) =	**7b**	435		
8	Total social security and Medicare taxes (add lines 6b, 6d, and 7b). Check here if wages are not subject to social security and/or Medicare tax . . . . ▶ ☐	**8**	2,295		
9	Adjustment of social security and Medicare taxes (see Instructions for required explanation) Sick Pay $_____ ± Fractions of Cents $_____ ± Other $_____ =	**9**	—		
10	Adjusted total of social security and Medicare taxes (line 8 as adjusted by line 9 — see instructions)	**10**	2,295		
11	**Total taxes** (add lines 5 and 10)	**11**	4,545		
12	Advance earned income credit (EIC) payments made to employees	**12**	—		
13	Net taxes (subtract line 12 from line 11). **If $1,000 or more, this must equal line 17, column (d) below (or line D of Schedule B (Form 941)**	**13**	4,545		
14	Total deposits for quarter, including overpayment applied from a prior quarter	**14**	4,545		
15	**Balance due** (subtract line 14 from line 13). See instructions	**15**	—		

Problem 3

2	Total wages and tips, plus other compensation			2	50,000	
3	Total income tax withheld from wages, tips, and sick pay			3	7,500	
4	Adjustment of withheld income tax for preceding quarters of calendar year			4	—	
5	Adjusted total of income tax withheld (line 3 as adjusted by line 4 — see instructions) . .			5	7,500	
6	Taxable social security wages	6a 44,000	× 12.4% (.124) =	6b	5,456	
	Taxable social security tips	6c 6,000	× 12.4% (.124) =	6d	744	
7	Taxable Medicare wages and tips	7a 50,000	× 2.9% (.029) =	7b	1,450	
8	Total social security and Medicare taxes (add lines 6b, 6d, and 7b). Check here if wages are not subject to social security and/or Medicare tax . . . . ▶ ☐			8	7,650	
9	Adjustment of social security and Medicare taxes (see Instructions for required explanation) Sick Pay $_____ ± Fractions of Cents $_____ ± Other $_____ =			9	—	
10	Adjusted total of social security and Medicare taxes (line 8 as adjusted by line 9 — see instructions)			10	7,650	
11	**Total taxes** (add lines 5 and 10)			11	15,150	
12	Advance earned income credit (EIC) payments made to employees			12	—	
13	Net taxes (subtract line 12 from line 11). **If $1,000 or more, this must equal line 17, column (d) below (or line D of Schedule B (Form 941)**			13	15,150	
14	Total deposits for quarter, including overpayment applied from a prior quarter			14	15,150	
15	**Balance due** (subtract line 14 from line 13). See instructions			15	—	

Problem 5

2	Total wages and tips, plus other compensation				2	29,500
3	Total income tax withheld from wages, tips, and sick pay				3	4,200
4	Adjustment of withheld income tax for preceding quarters of calendar year				4	—
5	Adjusted total of income tax withheld (line 3 as adjusted by line 4 — see instructions) . .				5	4,200
6	Taxable social security wages	6a	26,000	× 12.4% (.124) =	6b	3,224
	Taxable social security tips	6c	1,500	× 12.4% (.124) =	6d	186
7	Taxable Medicare wages and tips	7a	29,500	× 2.9% (.029) =	7b	856
8	Total social security and Medicare taxes (add lines 6b, 6d, and 7b). Check here if wages are not subject to social security and/or Medicare tax . . . . ▶ ☐				8	4,266
9	Adjustment of social security and Medicare taxes (see Instructions for required explanation) Sick Pay $_____ ± Fractions of Cents $_____ ± Other $_____ =				9	—
10	Adjusted total of social security and Medicare taxes (line 8 as adjusted by line 9 — see instructions)				10	4,266
11	**Total taxes** (add lines 5 and 10)				11	8,466
12	Advance earned income credit (EIC) payments made to employees				12	—
13	Net taxes (subtract line 12 from line 11). **If $1,000 or more, this must equal line 17, column (d) below (or line D of Schedule B (Form 941)**				13	8,466
14	Total deposits for quarter, including overpayment applied from a prior quarter				14	8,466
15	**Balance due** (subtract line 14 from line 13). See instructions				15	—

		1st Qtr.	2nd Qtr.	3rd Qtr.	4th Qtr.
7.	G	$19,500	$19,000	$19,000	$18,200
	H	21,000	25,000	29,000	12,900
	I	26,000	26,000	26,000	9,900

	Employee	Cumulative Earnings	4th Qtr. Earnings	Social Security Earnings	Medicare Earnings
9.	1	$ 66,000	$28,000	$21,900	$28,000
	2	55,000	29,000	29,000	29,000
	3	80,000	28,000	7,900	28,000

11.

Empl	1st Quarter Gross Wages	Inc. Tax	Soc. Sec. Wages	2nd Quarter Gross Wages	Inc. Tax	Soc. Sec. Wages	3rd Quarter Gross Wages	Inc. Tax	Soc. Sec. Wages	4th Quarter Gross Wages	Inc. Tax	Soc. Sec. Wages
A	$ 6,800	$ 980	$ 6,800	$ 7,400	$1,036	$ 7,400	$ 7,500	$ 1,050	$ 7,500	$ 7,900	$ 1,106	$ 7,900
B	17,600	1,960	17,600	20,800	2,072	20,800	18,000	2,100	18,000	19,300	2,242	19,300
C	18,800	2,492	18,800	18,600	2,604	18,600	20,700	2,818	20,700	20,500	2,775	20,500
D	29,000	4,060	29,000	29,500	3,990	29,500	28,900	4,046	28,900	28,800	4,032	500
Total	72,200	9,492	72,200	76,300	9,702	76,300	75,100	10,014	75,100	76,500	10,155	48,200

Problem 13

2	Total wages and tips, plus other compensation				2	75,100
3	Total income tax withheld from wages, tips, and sick pay				3	10,014
4	Adjustment of withheld income tax for preceding quarters of calendar year				4	—
5	Adjusted total of income tax withheld (line 3 as adjusted by line 4 — see instructions) . .				5	10,014
6	Taxable social security wages	6a	75,100	× 12.4% (.124) =	6b	9,312
	Taxable social security tips	6c	—	× 12.4% (.124) =	6d	—
7	Taxable Medicare wages and tips	7a	75,100	× 2.9% (.029) =	7b	2,178
8	Total social security and Medicare taxes (add lines 6b, 6d, and 7b). Check here if wages are not subject to social security and/or Medicare tax . . . . ▶ ☐				8	11,490
9	Adjustment of social security and Medicare taxes (see Instructions for required explanation) Sick Pay $_____ ± Fractions of Cents $_____ ± Other $_____ =				9	—
10	Adjusted total of social security and Medicare taxes (line 8 as adjusted by line 9 — see instructions)				10	11,490
11	**Total taxes** (add lines 5 and 10)				11	21,504
12	Advance earned income credit (EIC) payments made to employees				12	—
13	Net taxes (subtract line 12 from line 11). **If $1,000 or more, this must equal line 17, column (d) below (or line D of Schedule B (Form 941)**				13	21,504
14	Total deposits for quarter, including overpayment applied from a prior quarter				14	20,116
15	**Balance due** (subtract line 14 from line 13). See instructions				15	1,388

Problem 15

2	Total wages and tips, plus other compensation			**2**	76,500	
3	Total income tax withheld from wages, tips, and sick pay			**3**	10,155	
4	Adjustment of withheld income tax for preceding quarters of calendar year			**4**	—	
5	Adjusted total of income tax withheld (line 3 as adjusted by line 4 — see instructions) . .			**5**	10,155	
6	Taxable social security wages	**6a** 48,200	× 12.4% (.124) =	**6b**	5,977	
	Taxable social security tips	**6c** —	× 12.4% (.124) =	**6d**	—	
7	Taxable Medicare wages and tips	**7a** 76,500	× 2.9% (.029) =	**7b**	2,219	

8	Total social security and Medicare taxes (add lines 6b, 6d, and 7b). Check here if wages are not subject to social security and/or Medicare tax . . . . ▶ ☐	**8**	8,196	
9	Adjustment of social security and Medicare taxes (see Instructions for required explanation) Sick Pay $_____ ± Fractions of Cents $_____ ± Other $_____ =	**9**	—	
10	Adjusted total of social security and Medicare taxes (line 8 as adjusted by line 9 — see instructions)	**10**	8,196	
11	**Total taxes** (add lines 5 and 10)	**11**	18,351	
12	Advance earned income credit (EIC) payments made to employees	**12**	—	
13	Net taxes (subtract line 12 from line 11). **If $1,000 or more, this must equal line 17, column (d) below (or line D of Schedule B (Form 941)**	**13**	18,351	
14	Total deposits for quarter, including overpayment applied from a prior quarter	**14**	360	
15	**Balance due** (subtract line 14 from line 13). See instructions	**15**	—	

		1st Qtr.	2nd Qtr.	3rd Qtr.	4th Qtr.
17. (a)	G	$ 7,000	0	0	0
	H	7,000	0	0	0
	I	7,000	0	0	0
(b) Total		$21,000	0	0	0

(c) 1st Qtr. only
State   = .036 × $21,000 = $756
Federal = .008 × $21,000 = $168

		1st Qtr.	2nd Qtr.	3rd Qtr.	4th Qtr.
19. (a)	A	$ 6,800	200	0	0
	B	7,000	0	0	0
	C	7,000	0	0	0
	D	7,000	0	0	0
(b) Total		$21,800	200	0	0

(c) 1st Qtr.
State   = .036 × $27,800 = $1,001
Federal = .008 × $27,800 = $ 222

2nd Qtr.
State   = .036 × $200 = $7.20
Federal = .008 × $200 = $1.60

21. (a) Maximum Social Security taxable earnings:
$ 87,900 × .062 = $5,449.80
(b) $500,000 × .0145 = $7,250
(c) $ 7,000 × .008 = $ 56
(d) $ 7,000 × .054 = $ 378

Quick Practice - Straight line

1. (a) Annual depr. $= \dfrac{\$7,500 - \$300}{6 \text{ years}} = \$1,200/\text{yr.}$

   (b) Annual depr. $= \dfrac{\$5,800 - \$400}{9 \text{ years}} = \$600/\text{yr.}$

   (c) Annual depr. $= \dfrac{\$2,900 - \$700}{4 \text{ years}} = \$550/\text{yr.}$

Depreciation Schedule

Year	Book Value (End of Year)	Annual Depreciation	Accumulated Depreciation
0	$2,900	--	--
1	2,350	$550	$ 550
2	1,800	550	1,100
3	1,250	550	1,650
4	700	550	2,200

Quick Practice - Declining Balance

1. (a) Rate $= 2 \times \dfrac{1}{5} = \dfrac{2}{5}$ or 40%

   First year depr. $= 40\% \times \$1,900 = \$760$
   Book value $= \$1,900 - \$760 = \$1,140$

   Second year depr. $= 40\% \times \$1,140 = \$456$
   Book value $= \$1,140 - \$456 = \$684$

   Third year depr. $= 40\% \times \$684 = \$274$
   Book value $= \$684 - \$274 = \$410$

   Fourth year depr. $= 40\% \times \$410 = \$164$
   Book value $= \$410 - \$164 = \$246$

   Fifth year depr.   = book value end of fourth year = $246
   Book value $= 0$

1. (Continued)

(b) Rate = $2 \times \dfrac{1}{4} = \dfrac{2}{4} = \dfrac{1}{2}$ or 50%

First year depr. = 50% × $800 = $400
Book value   = $800 - $400 = $400

Second year depr. = 50% × $400 = $200
Book value   = $400 - $200 = $200

Third year depr. = 50% × $200 = $100
Book value   = $200 - $100 = $100

Fourth year depr. = book value end of third year = $100
Book value   = 0

(c) Rate = $2 \times \dfrac{1}{6} = \dfrac{2}{6} = \dfrac{1}{3}$

First year depr. = $\dfrac{1}{3}$ × $1,800 = $600

Book value   = $1,800 - $600 = $1,200

Second year depr. = $\dfrac{1}{3}$ × $1,200 = $400

Book value   = $1,200 - $400 = $800

Third year depr. = $\dfrac{1}{3}$ × $800 = $267

Book value   = $800 - $267 = $533

Fourth year depr. = $\dfrac{1}{3}$ × $533 = $178

Book value   = $533 - $178 = $355

Fifth year depr. = $\dfrac{1}{3}$ × $355 = $118

Book value   = $355 - $118 = $237

Sixth year depr. = $\dfrac{1}{3}$ × $237 = $79 but this amount

would bring book value above
residual value of $150.

1. (Continued)

Sixth year depr. = $237 - $150 = $87.

Depreciation Schedule

Year	Book Value (End of Year)	Annual Depreciation	Accumulated Depreciation
0	$1,800	--	--
1	1,200	$600	$ 600
2	800	400	1,000
3	533	267	1,267
4	355	178	1,445
5	237	118	1,563
6	150	87	1,650

Quick Practice - Units of Production

1. (a) Rate = $\dfrac{\$7,500 - \$300}{90,000 \text{ units}}$ = $0.08 per unit

Year	Units Produced	×	Unit Rate	=	Annual Depreciation
1	18,000	×	$0.08	=	$1,440
2	20,000	×	0.08	=	1,600
3	25,000	×	0.08	=	2,000
4	19,000	×	0.08	=	1,520
5	8,000	×	0.08	=	640

(b) Rate = $\dfrac{\$8,100 - \$600}{30,000 \text{ units}}$ = $0.25 per unit

Year	Units Produced	×	Unit Rate	=	Annual Depreciation
1	10,000	×	$0.25	=	$2,500
2	8,000	×	0.25	=	2,000
3	7,000	×	0.25	=	1,750
4	5,000	×	0.25	=	1,250

(c) Rate = $\dfrac{\$2,600 - \$200}{200,000 \text{ units}}$ = $0.012 per unit

1. (c) (Continued)

Year	Units Produced	×	Unit Rate	=	Annual Depreciation
1	35,000	×	$0.012	=	$420
2	44,000	×	0.012	=	528
3	40,000	×	0.012	=	480
4	33,000	×	0.012	=	396
5	28,000	×	0.012	=	336
6	20,000	×	0.012	=	240

Quick Practice - MACRS

1. (a)

Year	Annual Cost Recovery
1	$1,000
2	1,333
3	444
4	223*

*Rounded up to fully recover cost

(b)

Year	Annual Cost Recovery
1	$ 829
2	1,420
3	1,015
4	725
5	518
6	518
7	518
8	257*

*Rounded down to equal cost

Cost Recovery Schedule

(c) Year	Book Value (End of Year)	Annual Recovery	Accumulated Cost Recovery
0	$4,500	--	--
1	3,600	$ 900	$ 900
2	2,160	1,440	2,340
3	1,296	864	3,204
4	778	518	3,722
5	260	518	4,240
6	0	260*	4,500

*Rounded up to equal book value, fifth year

Section 1

1. (a) Cost $6,000   Annual depr. $= \dfrac{\$5,600}{7} = \$800$

      Scrap Value   − 400

      Depreciable base $5,600

Year	Book Value (End of Year)	Annual Depreciation	Accumulated Depreciation
0	$6,000	--	--
1	5,200	$800	$ 800
2	4,400	800	1,600
3	3,600	800	2,400
4	2,800	800	3,200
5	2,000	800	4,000
6	1,200	800	4,800
7	400	800	5,600

(b) Depreciation rate $= 2 \times \dfrac{1}{5} = \dfrac{2}{5}$ or 40%

First year depr. = 40% × $4,000 = $1,600
Book value   = $4,000 − $1,600 = $2,400

Second year depr. = 40% × $2,400 = $960
Book value   = $2,400 − $960 = $1,440

Third year depr. = 40% × $1,440 = $576
Book value   = $1,440 − $576 = $864

Fourth year depr. = 40% × $864 = $346
Book value   = $864 − $346 = $518
Fifth year depr. = $518 − $200 = $318

Year	Book Value (End of Year)	Annual Depreciation	Accumulated Depreciation
0	$4,000	--	--
1	2,400	$1,600	$1,600
2	1,440	960	2,560
3	864	576	3,136
4	518	346	3,482
5	200	318*	3,800

*Adjusted to equal fifth year book value

1. (Continued)

(c)   MACRS Cost Recovery Schedule - 3 yr. class

Year	Book Value (End of Year)	Annual Cost Recovery	Accumulated Cost Recovery
0	$1,400	--	--
1	933	$467	$ 467
2	311	622	1,089
3	104	207	1,296
4	0	104	1,400

(d)   MACRS Cost Recovery Schedule - 7 yr. class

Year	Book Value (End of Year)	Annual Cost Recovery	Accumulated Cost Recovery
0	$3,600	--	--
1	3,086	$514	$ 514
2	2,204	882	1,396
3	1,574	630	2,026
4	1,124	450	2,476
5	803	321	2,797
6	482	321	3,118
7	161	321	3,439
8	0	161	3,600

3. Cost $6,800   Annual depr. $= \dfrac{\$6,300}{5 \text{ yrs.}} = \$1,260$
   Residual value $-$ 500
   Depreciable base $6,300

Depreciation Schedule

Year	Book Value (End of Year)	Annual Depreciation	Accumulated Depreciation
0	$6,800	--	--
1	5,540	$1,260	$1,260
2	4,280	1,260	2,520
3	3,020	1,260	3,780
4	1,760	1,260	5,040
5	500	1,260	6,300

5. Cost $2,700   Annual depr. $= \dfrac{\$2,400}{6} = \$400$
   Salvage value $-$ 300
   Depreciable base $2,400

7. Cost $38,000   Annual depr. $= \dfrac{\$32,000}{10} = \$3,200$
   Residual value $-$ 6,000
   Depreciable base $32,000

9. Depreciation rate $= 2 \times \dfrac{1}{6} = \dfrac{2}{6} = \dfrac{1}{3}$

First year depr. $= \dfrac{1}{3} \times \$12,000 = \$4,000$

Book value $= \$12,000 - \$4,000 = \$8,000$

Second year depr. $= \dfrac{1}{3} \times \$8,000 = \$2,667$

Book value $= \$8,000 - \$2,667 = \$5,333$

Third year depr. $= \dfrac{1}{3} \times \$5,333 = \$1,778$

Book value $= \$5,333 - \$1,778 = \$3,555$

Fourth year depr. $= \dfrac{1}{3} \times \$3,555 = \$1,185$

Book value $= \$3,555 - \$1,185 = \$2,370$

Fifth year depr. $= \dfrac{1}{3} \times \$2,370 = \$790$

Book value $= \$2,370 - \$790 = \$1,580$

Sixth year depr. $= \$1,580 - \$800 = \$780$

Depreciation Schedule

Year	Book Value (End of Year)	Annual Depreciation	Accumulated Depreciation
0	$12,000	--	--
1	8,000	$4,000	$ 4,000
2	5,333	2,667	6,667
3	3,555	1,778	8,445
4	2,370	1,185	9,630
5	1,580	790	10,420
6	800	780*	11,200

*Adjusted so book value equals $800 (residual value)

11. Depreciation rate $= 2 \times \dfrac{1}{5} = \dfrac{2}{5} = 40\%$

Depreciation Schedule

Year	Book Value (End of Year)	Annual Depreciation	Accumulated Depreciation
0	$6,000	--	--
1	3,600	$2,400	$2,400
2	2,160	1,440	3,840
3	1,296	864	4,704
4	778	518	5,222
5	0	778*	6,000

*Adjusted so book value equals $0 (scrap value)

13. Depreciation rate $= 2 \times \dfrac{1}{6} = \dfrac{2}{6} = \dfrac{1}{3}$ or 33 1/3%

Depreciation Schedule

Year	Book Value (End of Year)	Annual Depreciation	Accumulated Depreciation
0	$800	--	--
1	533	$267	$267
2	355	178	445
3	237	118	563
4	158	79	642
5	105	53	695
6	100	5*	700

*Adjusted so book value equals $100 (residual value)

15.

Cost Recovery Schedule - 5 yr. class

Year	Book Value (End of Year)	Annual Cost Recovery	Accumulated Cost Recovery
0	$28,600	--	--
1	22,880	$5,720	$5,720
2	13,728	9,152	14,872
3	8,237	5,491	20,363
4	4,942	3,295	23,658
5	1,647	3,295	26,953
6	0	1,647	28,600

17.

Year	Cost	×	Cost Recovery Factor	=	Cost Recovery (Rounded)
1	$800,000	×	0.030423	=	$24,338
2	800,000	×	0.031746	=	25,397
3	800,000	×	0.031746	=	25,397

19.

Year	27.5-Year $600,000	5-Year $25,000	Total Cost Recovery
1	$20,909	$ 5,000	$25,909
2	21,818	8,000	29,818
3	21,818	4,800	26,618
	$64,545	$17,800	$82,345

21. (a)

Cost	$3,000
Salvage value	− 900
Depreciable base	$2,100

Rate = $\dfrac{\$2,100}{300,000 \text{ units}}$

= $0.007 per unit

Year	Annual Units	×	Unit Rate	=	Annual Depreciation
1	75,000	×	$0.007	=	$ 525
2	82,000	×	0.007	=	574
3	76,000	×	0.007	=	532
4	67,000	×	0.007	=	469
	300,000				$2,100

(b)

Cost	$5,700
Salvage value	− 300
Depreciable base	$5,400

Rate = $\dfrac{\$5,400}{300,000 \text{ units}}$

= $0.018 per unit

Year	Annual Units	×	Unit Rate	=	Annual Depreciation
1	70,000	×	$0.018	=	$1,260
2	80,000	×	0.018	=	1,440
3	88,000	×	0.018	=	1,584
4	62,000	×	0.018	=	1,116
	300,000				$5,400

(c)

Cost	$6,800
Salvage value	− 500
Depreciable base	$6,300

Rate = $\dfrac{\$6,300}{300,000 \text{ units}}$

= $0.021 per unit

21. (c) (Continued)

Year	Annual Units	×	Unit Rate	=	Annual Depreciation
1	56,000	×	$0.021	=	$1,176
2	58,000	×	0.021	=	1,218
3	60,000	×	0.021	=	1,260
4	55,000	×	0.021	=	1,155
5	40,000	×	0.021	=	840
6	31,000	×	0.021	=	651
	300,000				$6,300

23. Straight-line method:

Cost $4,200
Residual value   -   200
Depreciable base $4,000

$$\text{Annual Depr.} = \frac{\$4,000}{5} = \$800$$

Declining-balance method:

$$\text{Rate} = 2 \times \frac{1}{5} = \frac{2}{5} \text{ or } 40\%$$

First year depr. = 40% × $4,200 = $1,680
Book value   = $4,200 - $1,680 = $2,520

Second year depr. = 40% × $2,520 = $1,008
Book value   = $2,520 - $1,008 = $1,512

Third year depr. = 40% × $1,512 = $605
Book value   = $1,512 - $605 = $907

MACRS method:

First year cost recovery = 0.2   × $4,200 = $ 840
Second year cost recovery = 0.32 × $4,200 = $1,344
Third year cost recovery = 0.192 × $4,200 = $ 806

Year	Straight Line	Declining Balance	MACRS
1	$800	$1,680	$ 840
2	800	1,008	1,344
3	800	605	806

Section 2

1. (a) MACRS Cost Recovery Schedule - 5 yr. class

Year	Book Value (End of Year)	Annual Cost Recovery	Accumulated Cost Recovery
0	$9,000	--	--
1	7,200	$1,800	$1,800
2	4,320	2,880	4,680
3	2,592	1,728	6,408
4	1,555	1,037	7,445
5	518	1,037	8,482
6	0	518	9,000

(b)
Cost $4,900

Residual value   − 400

Depreciable base $4,500

Straight-line method:

$$\frac{\$4,500}{5} = \$900 \text{ per full year}$$

First year $= \frac{1}{4}$ of a year; $\frac{1}{4} \times \$900 = \$225$

Sixth year $= \frac{3}{4}$ of a year; $\frac{3}{4} \times \$900 = \$675$

Depreciation Schedule

Year	Book Value (End of Year)	Annual Depreciation	Accumulated Depreciation
0	$4,900	--	--
1	4,675	$225	$ 225
2	3,775	900	1,125
3	2,875	900	2,025
4	1,975	900	2,925
5	1,075	900	3,825
6	400	675	4,500

(c) Depreciable base $= \$2,600$

Depreciation rate $= 2 \times \frac{1}{4} = \frac{2}{4} = \frac{1}{2}$ or 50%

First year $= \frac{3}{4}$ of a year; $\frac{3}{4} \times 50\% \times \$2,600 = \$975$

1. (c) (Continued)

Depreciation Schedule

Year	Book Value (End of Year)	Annual Depreciation	Accumulated Depreciation
0	$2,600	--	--
1	1,625	$975	$ 975
2	812	813	1,788
3	406	406	2,194
4	203	203	2,397
5	0	203*	2,600

*Adjusted to bring book value to $0

3. Cost $5,300 Straight-line method:
    Trade-in value − 400
    Depreciable base $4,900 $\dfrac{\$4,900}{7}$ = $700 per full year

First year = $\dfrac{2}{12}$ of a year; $\dfrac{2}{12}$ × $700 = $117

Eighth year = $\dfrac{10}{12}$ of a year; $\dfrac{10}{12}$ × $600 = $583

Depreciation Schedule

Year	Book Value (End of Year)	Annual Depreciation	Accumulated Depreciation
0	$5,300	--	--
1	5,183	$117	$ 117
2	4,483	700	817
3	3,783	700	1,517
4	3,083	700	2,217
5	2,383	700	2,917
6	1,683	700	3,617
7	983	700	4,317
8	400	583	4,900

5. Declining-balance rate: $2 \times \dfrac{1}{4} = \dfrac{2}{4} = \dfrac{1}{2}$ or 50%

First year = $\dfrac{4}{12}$ of a year; $\dfrac{4}{12} \times \dfrac{1}{2}$ × $900 = $150

5. (Continued)

| | Depreciation Schedule | | |
| | Book Value | Annual | Accumulated |
Year	(End of Year)	Depreciation	Depreciation
0	$900	--	--
1	750	$150	$150
2	375	375	525
3	187	188	713
4	93	94	807
5	50	43*	850

*Adjusted

Section 3

1. (a)

Dept.	Ratio of floor space	Overhead charge
#1	$\dfrac{1,800}{5,500} = \dfrac{18}{55}$	$19,250 \times \dfrac{18}{55} = \$\ 6,300$
2	$\dfrac{1,000}{5,500} = \dfrac{2}{11}$	$19,250 \times \dfrac{2}{11} = 3,500$
3	$\dfrac{1,200}{5,500} = \dfrac{12}{55}$	$19,250 \times \dfrac{12}{55} = 4,200$
4	$\dfrac{1,500}{5,500} = \dfrac{3}{11}$	$19,250 \times \dfrac{3}{11} = \underline{5,250}$

Total overhead = $19,250

(b)

Dept.	Ratio of floor space	Overhead charge
#16	$\dfrac{500}{2,000} = \dfrac{1}{4}$	$32,000 \times \dfrac{1}{4} = \$\ 8,000$
17	$\dfrac{300}{2,000} = \dfrac{3}{20}$	$32,000 \times \dfrac{3}{20} = 4,800$
18	$\dfrac{400}{2,000} = \dfrac{1}{5}$	$32,000 \times \dfrac{1}{5} = 6,400$
19	$\dfrac{800}{2,000} = \dfrac{2}{5}$	$32,000 \times \dfrac{2}{5} = \underline{12,800}$

Total overhead = $32,000

3. (a)

Dept.	Ratio of net sales	Overhead charge
W	$\dfrac{30,000}{70,000} = \dfrac{3}{7}$	$\$56,000 \times \dfrac{3}{7} = \$24,000$
X	$\dfrac{15,000}{70,000} = \dfrac{3}{14}$	$\$56,000 \times \dfrac{3}{14} = 12,000$
Y	$\dfrac{14,000}{70,000} = \dfrac{1}{5}$	$\$56,000 \times \dfrac{1}{5} = 11,200$
Z	$\dfrac{11,000}{70,000} = \dfrac{11}{70}$	$\$56,000 \times \dfrac{11}{70} = \underline{\quad 8,800}$

Total overhead = $56,000

(b)

Dept.	Ratio of net sales	Overhead charge
AA	$\dfrac{20,000}{72,000} = \dfrac{5}{18}$	$\$54,000 \times \dfrac{5}{18} = \$15,000$
BB	$\dfrac{22,000}{72,000} = \dfrac{11}{36}$	$\$54,000 \times \dfrac{11}{36} = 16,500$
CC	$\dfrac{14,000}{72,000} = \dfrac{7}{36}$	$\$54,000 \times \dfrac{7}{36} = 10,500$
DD	$\dfrac{16,000}{72,000} = \dfrac{2}{9}$	$\$54,000 \times \dfrac{2}{9} = \underline{\quad 12,000}$

Total overhead = $54,000

5. (a)

Dept.	Ratio of # employees	Overhead charge
#22	$\dfrac{40}{100} = \dfrac{2}{5}$	$\$18,000 \times \dfrac{2}{5} = \$\ 7,200$
33	$\dfrac{10}{100} = \dfrac{1}{10}$	$\$18,000 \times \dfrac{1}{10} = 1,800$
44	$\dfrac{30}{100} = \dfrac{3}{10}$	$\$18,000 \times \dfrac{3}{10} = 5,400$
55	$\dfrac{20}{100} = \dfrac{1}{5}$	$\$18,000 \times \dfrac{1}{5} = \underline{\quad 3,600}$

Total overhead = $18,000

5. (Continued)

(b)

Dept.	Ratio of # employees	Overhead charge
#10	$\dfrac{2}{15}$	$\$30,000 \times \dfrac{2}{15} = \$ \ 4,000$
11	$\dfrac{4}{15}$	$\$30,000 \times \dfrac{4}{15} = \ \ 8,000$
12	$\dfrac{3}{15} = \dfrac{1}{5}$	$\$30,000 \times \dfrac{1}{5} = \ \ 6,000$
13	$\dfrac{6}{15} = \dfrac{2}{5}$	$\$30,000 \times \dfrac{2}{5} = \ \underline{12,000}$

Total overhead = $30,000

7.

Dept.	Ratio of Floor space	Overhead charge
Accounting	$\dfrac{200}{1,500} = \dfrac{2}{15}$	$\$75,000 \times \dfrac{2}{15} = \$10,000$
General office	$\dfrac{300}{1,500} = \dfrac{1}{5}$	$\$75,000 \times \dfrac{1}{5} = \ 15,000$
Marketing	$\dfrac{400}{1,500} = \dfrac{4}{15}$	$\$75,000 \times \dfrac{4}{15} = \ 20,000$
Service/repairs	$\dfrac{600}{1,500} = \dfrac{2}{5}$	$\$75,000 \times \dfrac{2}{5} = \ \underline{30,000}$

Total overhead = $75,000

9.

Dept.	Ratio of net sales	Overhead charge
Biographies	$\dfrac{60,000}{200,000} = \dfrac{3}{10}$	$\$120,000 \times \dfrac{3}{10} = \$36,000$
Adv./Mysteries	$\dfrac{44,000}{200,000} = \dfrac{11}{50}$	$\$120,000 \times \dfrac{11}{50} = \ 26,400$
Computer/Software	$\dfrac{42,000}{200,000} = \dfrac{21}{100}$	$\$120,000 \times \dfrac{21}{100} = \ 25,200$
Romance	$\dfrac{38,000}{200,000} = \dfrac{19}{100}$	$\$120,000 \times \dfrac{19}{100} = \ 22,800$
Sports	$\dfrac{16,000}{200,000} = \dfrac{2}{25}$	$\$120,000 \times \dfrac{2}{25} = \ \underline{9,600}$

Total overhead = $120,000

11.

Dept.	Ratio of # employees	Overhead charge
Lawn/Garden	$\dfrac{7}{30}$	$\$90,000 \times \dfrac{7}{30} = \$21,000$
Appliances	$\dfrac{8}{30} = \dfrac{4}{15}$	$\$90,000 \times \dfrac{4}{15} = 24,000$
Automotive	$\dfrac{5}{30} = \dfrac{1}{6}$	$\$90,000 \times \dfrac{1}{6} = 15,000$
Bldg. supplies	$\dfrac{6}{30} = \dfrac{1}{5}$	$\$90,000 \times \dfrac{1}{5} = 18,000$
Paint	$\dfrac{4}{30} = \dfrac{2}{15}$	$\$90,000 \times \dfrac{2}{15} = \underline{12,000}$

$$\text{Total overhead} = \$90,000$$

Sections 1 & 2

1.

<table>
<tr><td colspan="5" align="center">Zatron Corporation
Income Statement
For Year Ending Dec. 31, 20X2</td></tr>
<tr><td>Income from sales:</td><td></td><td></td><td></td><td></td></tr>
<tr><td>Sales</td><td>$ 520,000</td><td></td><td>104.0%</td><td></td></tr>
<tr><td>Sales discount</td><td>20,000</td><td></td><td>4.0%</td><td></td></tr>
<tr><td>Net sales</td><td></td><td>$ 500,000</td><td></td><td>100.0%</td></tr>
<tr><td>Cost of goods sold:</td><td></td><td></td><td></td><td></td></tr>
<tr><td>Inventory, Jan. 1</td><td>$ 82,000</td><td></td><td></td><td></td></tr>
<tr><td>Net purchases</td><td>150,000</td><td></td><td></td><td></td></tr>
<tr><td>Goods avail. for sale</td><td>232,000</td><td></td><td></td><td></td></tr>
<tr><td>Inventory, Dec. 31</td><td>96,000</td><td></td><td></td><td></td></tr>
<tr><td>Cost of goods sold</td><td></td><td>$ 136,000</td><td></td><td>27.2%</td></tr>
<tr><td>Gross profit</td><td></td><td>$ 364,000</td><td></td><td>72.8%</td></tr>
<tr><td>Operating Expenses</td><td></td><td></td><td></td><td></td></tr>
<tr><td>Salaries</td><td>$ 196,000</td><td></td><td>39.2%</td><td></td></tr>
<tr><td>Depreciation</td><td>21,500</td><td></td><td>4.3%</td><td></td></tr>
<tr><td>Utilities</td><td>18,000</td><td></td><td>3.6%</td><td></td></tr>
<tr><td>Maintenance</td><td>16,500</td><td></td><td>3.3%</td><td></td></tr>
<tr><td>Advertising</td><td>10,000</td><td></td><td>2.0%</td><td></td></tr>
<tr><td>Insurance</td><td>8,500</td><td></td><td>1.7%</td><td></td></tr>
<tr><td>Office supplies</td><td>8,000</td><td></td><td>1.6%</td><td></td></tr>
<tr><td>Miscellaneous</td><td>3,200</td><td></td><td>0.6%</td><td></td></tr>
<tr><td>Total expenses</td><td></td><td>$ 281,700</td><td></td><td>56.3%</td></tr>
<tr><td>Net income from operations</td><td></td><td>$ 82,300</td><td></td><td>16.5%</td></tr>
<tr><td>Income taxes</td><td></td><td>14,000</td><td></td><td>2.8%</td></tr>
<tr><td>Net income after taxes</td><td></td><td>$ 68,300</td><td></td><td>13.7%</td></tr>
</table>

3.

Madison & Co. Balance Sheet, December 20X1				%	%
Assets					
Current Assets:					
Cash		$ 27,000		4.7%	
Accounts receivable		32,000		5.6%	
Notes receivable		20,000		3.5%	
Inventory		45,000		7.8%	
Total current assets			$ 124,000		21.6%
Plant assets:					
Building	$ 300,000				
Less: Accum. Depreciation	30,000				
Net building		$ 270,000		47.0%	
Truck	38,000				
Less: Accum. Depreciation	18,000				
Net truck		$ 20,000		3.5%	
Land		160,000			
Total plant assets			$ 450,000		78.4%
Total assets			$ 574,000		100.0%
Liabilities and Owner's Equity					
Current liabilities:					
Accounts payable		$ 34,000		5.9%	
Notes payable		46,000		8.0%	
Total current liabilities			$ 80,000		13.9%
Long-term liabilities:					
Mortgage			229,000		39.9%
Total liabilities			$ 309,000		53.8%
Owner's equity			265,000		46.2%
Total liabilities and owner's equity			$ 574,000		100.0%

5.

Sabin Automotive Inc. Comparative Income Statement For Months of July and August, 20XX						
			Increase or (Decrease)		Percent of Net Sales	
	August	July	Amount	Percent	August	July
Income:						
Net sales	$ 325,000	$ 300,000				
Cost of goods sold:						
Inventory, Jan. 1	$ 90,000	82,000				
Purchases	135,000	150,000				
Goods avail. for sale						
Inventory, Dec. 31	81,000	90,000				
Cost of goods sold						
Gross profit						
Expenses:						
Salaries	$ 75,000	60,000				
Rent	44,000	42,500				
Advertising	6,000	6,000				
Depreciation	5,000	5,500				
Utilities	2,900	2,800				
Miscellaneous	3,500	2,200				
Total expenses						
Net income						

7.

Davie & Associates Comparative Balance Sheet December 31, 20X2 and 20X1						
			Increase or (Decrease)		Percent of Net Sales	
	20X2	20X1	Amount	Percent	20X2	20X1
Assets						
Current assets:						
Cash	$ 27,000	$ 33,000	$ (6,000)	(18.2%)	5.8%	6.8%
Accounts receivable	52,000	48,000	4,000	8.3%	11.3%	9.9%
Inventory	83,000	82,000	1,000	1.2%	18.0%	16.9%
Total current assets	162,000	163,000	(1,000)	(0.6%)	35.1%	33.6%
Total plant assets	300,000	322,000	(22,000)	(6.8%)	64.9%	66.4%
Total assets	462,000	485,000	(23,000)	(4.7%)	100.0%	100.0%
liabilities and Equity						
Current liabilities	79,000	85,000	(6,000)	(7.1%)	17.1%	17.5%
Long-term liabilities	150,000	172,000	(22,000)	(12.8%)	32.5%	35.5%
Total liabilities	229,000	257,000	(28,000)	(10.9%)	49.6%	53.0%
Stockholders's equity:						
Preferred stock	45,000	45,000		0.0%	9.7%	9.3%
Common stock	80,000	78,000	2,000	2.6%	17.3%	16.1%
Retained earnings	108,000	105,000	3,000	2.9%	23.4%	21.6%
Total equity	233,000	228,000	5,000	2.2%	50.4%	47.0%
Total liabilities and equity	$ 462,000	$ 485,000	(23,000)	(4.7%)	100.0%	100.0%

9. (a)   Working capital ratio:

$$\frac{\text{Current assets}}{\text{Current liabilities}} = \frac{\$124,000}{\$80,000} = 1.6 \text{ to } 1$$

(b) Acid-test ratio:

$$\frac{\text{Quick assets}}{\text{Current liabilities}} = \frac{\$79,000}{\$80,000} = 1.0 \text{ to } 1$$

(c) Ratio of net sales to net working capital:

$$\frac{\text{Net sales}}{\text{Current assets} - \text{Current liabilities}} = \frac{\$465,000}{\$124,000 - \$80,000}$$

$$= 10.6 \text{ to } 1$$

(d) Book value of stock:

$$\frac{\text{Owners' equity}}{\text{Number of shares of stock}} = \frac{\$265,000}{15,000} = \$17.67$$

11.   (a)   Working capital ratio:

$$\frac{\text{Current assets}}{\text{Current liabilities}} = \frac{\$162,000}{\$79,000} = 2.1 \text{ to } 1$$

(b) Acid-test ratio:

$$\frac{\text{Quick assets}}{\text{Current liabilities}} = \frac{\$79,000}{\$79,000} = 1 \text{ to } 1$$

(c) Book Value of Stock:

$$\frac{\text{Owners' Equity}}{\text{Number of Shares of Stock}} = \frac{\$233,000}{\$120,000} = \$1.94$$

Section 3

1.   Inventory
          at retail:
          $146,000   Average inventory = $\frac{\$630,000}{4}$ = $157,500
           161,000   (at retail)
           153,000
           170,000
          $630,000

3. (a) Cost $ 91,000   (b) Inventory
    +35% of cost + 31,850   at retail:
    Retail $122,850 $122,850
                                     100,000
                                     104,000
                                    126,150
                                  $453,000

$$\text{Average inventory (at retail)} = \frac{\$453,000}{4} = \$113,250$$

5. $$\text{Inventory turnover} = \frac{\text{Net sales}}{\text{Average inventory at retail}}$$
$$= \frac{\$641,800}{\$157,500}$$
$$= 4.1 \text{ times}$$

7. From problems 1 and 5:
Average inventory at retail = $157,500
Cost of goods sold = $481,000

(a) Cost = 60% of retail
= 0.60($157,500)
C = $94,500

(b) $$\text{Turnover} = \frac{\text{Cost of goods sold}}{\text{Average inventory at cost}}$$
$$= \frac{\$481,000}{\$94,500}$$
$$= 5.1 \text{ times}$$

9. Inventory at retail:
$116,000
124,000
125,000
126,000
$491,000

$$\text{Average inventory (at retail)} = \frac{\$491,000}{4} = \$122,750$$

$$\text{Turnover} = \frac{\text{Net sales}}{\text{Average inventory at retail}} = \frac{\$392,800}{\$122,750} = 3.2 \text{ times}$$

11. From problem 9:
    Average inventory at retail = $122,750;
    Cost of goods sold = $250,410

(a) Cost = 60% of retail
          = 0.60($122,750)
      C = $73,650

    Average inventory at
    cost = $73,650

(b)
$$\frac{\text{Cost of goods sold}}{\text{Average inventory at cost}}$$

$$= \frac{\$250,410}{\$73,650} = 3.4 \text{ times}$$

Section 4

1. (a) 10 @ $5.10 = $ 51
        30 @ 5.20 = 156
        20 @ 5.25 = 105
        60 units $312

    (1) Weighted average:

    $$\frac{\$312}{60} = \$5.20 \text{ average cost}$$

    25 @ $5.20 = $130 inventory value

    (2) FIFO:
        20 @ $5.25 = $105
         5 @ 5.20 =   26
        25 units $131 inventory value

    (3) LIFO:
        10 @ $5.10 = $ 51
        15 @ 5.20 =   78
        25 units $129 inventory value

(b) 25 @ $7.40 = $185.00
     45 @ 7.50 = 337.50
     30 @ 7.65 = 229.50
    100 units $752.00

    (1) Weighted average:

    $$\frac{\$752}{100} = \$7.52 \text{ average cost}$$

    40 @ $7.52 = $300.80 inventory value

    (2) FIFO:
        30 @ $7.65 = $229.50
        10 @ 7.50 =   75.00
        40 units $304.50 inventory value

1. (b) (Continued)

        (3) <u>LIFO</u>:
            25 @ $7.40 = $185.00
            <u>15</u> @   7.50 =   <u>112.50</u>
            40 units $297.50 inventory value

3.   31 @ $116 = $ 3,596
     32 @ 118 =   3,776
     40 @ 120 =   4,800
     44 @ 121 =   5,324
     <u>28</u> @ 125 =   <u>3,500</u>
  175 units $20,996

(a) <u>Weighted average</u>:

$$\frac{\$20,996}{175} = \$119.98 \text{ average cost}$$

10 @ $119.98 = $1,199.80 inventory value

(b) <u>FIFO</u>:
10 @ $125 = $1,250 inventory value

(c) <u>LIFO</u>:
10 @ $116 = $1,160 inventory value

5.   40 @ $35 = $1,400
    80 @ 34 = 2,720
    60 @ 32 = 1,920
  <u> 70</u> @ 30 = <u>2,100</u>
  250 units $8,140

(a) <u>Weighted average</u>:

$$\frac{\$8,140}{250} = \$32.56 \text{ average cost}$$

60 @ $32.56 = $1,953.60 inventory cost

(b) <u>FIFO</u>:
60 @ $30 = $1,800 inventory cost

(c) <u>LIFO</u>:
40 @ $35 = $1,400
<u>20</u> @ 34 =   <u>680</u>
60 units $2,080 inventory cost

7.

	Wt. Avg.		FIFO		LIFO
Net sales		$40,000.00		$40,000	$40,000
Cost of goods sold:					
Beg. inventory	$ 0		$ 0		$ 0
Purchases	20,996.00		20,996		20,996
Goods available	$20,996.00		$20,996		$20,996
Endg. inventory	1,199.80		1,250		1,160
Cost of goods		19,796.20		19,746	19,836
Gross profit		$20,203.80		$20,254	$20,164
% gross profit		50.5%		50.6%	50.4%

Section 1

1. (a) $\dfrac{\$63,000}{450,000 \text{ shares}}$ = $0.14 dividend per share common

   (b) 6% × $100 par value = $6 dividend per share preferred

   $6 × 10,000 shares = $60,000

   $100,000 Total dividend
   $\underline{-\ 60,000}$ Total preferred dividend
   $ 40,000 Total common dividend

   $\dfrac{\$40,000}{500,000 \text{ shares}}$ = $0.08 dividend per share common

   (c) 5% × $50 par value × 2 years = $5 dividend per share
   preferred

   $5 × 40,000 shares = $200,000

   $220,000 Total dividend
   $\underline{-200,000}$ Total preferred dividend
   $ 20,000 Total common dividend

   $\dfrac{\$20,000}{200,000 \text{ shares}}$ = $0.10 dividend per share common

3. (a) <u>Dividend per share</u>:
   7% × $100 par value = $7 dividend per share preferred

   $7 × 6,000 shares = $42,000

   $450,000 Total dividend
   $\underline{-\ 42,000}$ Total preferred dividend
   $408,000 Total common dividend

3. (a) (Continued)

$$\frac{\$408,000}{300,000 \text{ shares}} = \$1.36 \text{ dividend per share common}$$

Earnings per share:

$$\frac{\$900,000 - \$42,000}{300,000 \text{ shares}} = \$2.86 \text{ earnings per share}$$

(b) Dividend per share:
8% × $50 par value = $4 dividend per share preferred

$4 × 20,000 shares = $80,000

$150,000 Total dividend
- 80,000 Total preferred dividend
$ 70,000 Total common dividend

$$\frac{\$70,000}{10,000 \text{ shares}} = \$7 \text{ dividend per share common}$$

Earnings per share:

$$\frac{\$200,000 - \$80,000}{10,000 \text{ shares}} = \$12 \text{ earnings per share}$$

5. $$\frac{\text{Total dividend}}{\text{Total shares}} = \frac{\$136,000}{680,000} = \$0.20 \text{ dividend per share common}$$

7. 7% × $100 par value = $7 dividend per share preferred

$7 × 20,000 shares = $140,000

$175,000 Total dividend
-140,000 Total preferred dividend
$ 35,000 Total common dividend

$$\frac{\$35,000}{100,000 \text{ shares}} = \$0.35 \text{ dividend per share common}$$

9. 8% × $50 par value = $4 dividend per share preferred

   $4 × 1,500 shares = $6,000

   $10,000 Total dividend
   $\underline{-\ 6,000}$ Total preferred dividend
   $ 4,000 Total common dividend

   $\dfrac{\$4,000}{500 \text{ shares}}$ = $8 dividend per share common

11. 6% × $100 par value = $6

    $6 × 65,000 shares = $390,000 but only $325,000 total dividend

    $\dfrac{\$325,000}{65,000 \text{ shares}}$ = $5 dividend per share preferred

                 0 dividend per share common

13. 7% × $50 par value × 2 years = $7 dividend per share
                                    preferred

    $7 × 4,000 shares = $28,000

    $77,000 Total dividend
    $\underline{-28,000}$ Total preferred dividend
    $49,000 Total common dividend

    $\dfrac{\$49,000}{5,000 \text{ shares}}$ = $9.80 dividend per share common

15. 8% × $50 × 3 years = $12 dividend per share preferred

    $12 × 3,000 shares = $36,000

    $40,000 Total dividend
    $\underline{-36,000}$ Total preferred dividend
    $ 4,000 Total common dividend

    $\dfrac{\$4,000}{100,000 \text{ shares}}$ = $0.04 dividend per share common

17. 5% × $100 par value × 2 years = $10 dividend per share
                                        preferred

$10 × 10,000 shares = $100,000

$130,000 Total dividend
-100,000 Total preferred dividend
$ 30,000 Total common dividend

$$\frac{\$30,000}{6,000 \text{ shares}} = \$5 \text{ dividend per share common}$$

19. 6% × $50 par value = $3 dividend per share preferred (stock
                                is non-cumulative)

$3 × 10,000 shares = $30,000

$45,000 Total dividend
-30,000 Total preferred dividend
$15,000 Total common dividend

$$\frac{\$15,000}{150,000 \text{ shares}} = \$0.10 \text{ dividend per share common}$$

21. (a) 6% × $50 par value = $3.00 dividend per share preferred

$3.00 × 4,500 shares = $13,500

$90,000 Total dividend
-13,500 Total preferred dividend
$76,500 Total common dividend

$$\frac{\$76,500}{10,000 \text{ shares}} = \$7.65 \text{ dividend per share common}$$

(b) $$\frac{\$150,000 - \$13,500}{10,000 \text{ shares}} = \$13.65 \text{ earnings per share}$$

Section 2

1. (a) 1 + 3 = 4 shares

$$K = \frac{1}{4} \times \$600,000 = \$150,000$$

$$L = \frac{3}{4} \times \$600,000 = \underline{\$450,000}$$
$$\$600,000$$

1. (Continued)

(b) $A = \dfrac{\$50,000}{\$180,000} = \dfrac{5}{18}$; $\dfrac{5}{18} \times \$216,000 = \$\ 60,000$

$B = \dfrac{\$60,000}{\$180,000} = \dfrac{1}{3}$; $\dfrac{1}{3} \times \$216,000 = \$\ 72,000$

$C = \dfrac{\$70,000}{\$180,000} = \dfrac{7}{18}$; $\dfrac{7}{18} \times \$216,000 = \underline{\$\ 84,000}$
$\qquad\qquad\qquad\qquad\qquad\qquad\qquad\qquad\quad \$216,000$

(c)

	S	T	U
Investment	$300,000	$350,000	$150,000
Interest rate	× 6%	× 6%	× 6%
Interest	$ 18,000	$ 21,000	$ 9,000

Total interest = $18,000 + $21,000 + $9,000 = $48,000

$160,000 Total profit
– 48,000 Interest
$112,000 To be shared in a ratio of 2:2:1

2 + 2 + 1 = 5 shares

$S = \dfrac{2}{5} \times \$112,000 = \$44,800$

$T = \dfrac{2}{5} \times \$112,000 = \$44,800$

$U = \dfrac{1}{5} \times \$112,000 = \$22,400$

SUMMARY

	S	T	U	Check
Interest	$18,000	$21,000	$ 9,000	$62,800
Ratio	44,800	44,800	22,400	65,800
Total	$62,800	$65,800	$31,400	31,400
				$160,000

(d)

	V	W	X
Investment	$300,000	$280,000	$240,000
Interest rate	× 5%	× 5%	× 5%
Interest	$ 15,000	$ 14,000	$ 12,000

Total interest = $15,000 + $14,000 + $12,000 = $41,000

1. (Continued)

$437,000 Total profit
- 41,000 Interest
-150,000 Salary to V
$246,000 To be shared in ratio of investments

$$V = \frac{\$300,000}{\$820,000} = \frac{15}{41}; \quad \frac{15}{41} \times \$246,000 = \$90,000$$

$$W = \frac{\$280,000}{\$820,000} = \frac{14}{41}; \quad \frac{14}{41} \times \$246,000 = \$84,000$$

$$X = \frac{\$240,000}{\$820,000} = \frac{12}{41}; \quad \frac{12}{41} \times \$246,000 = \frac{\$72,000}{\$246,000}$$

SUMMARY

	V	W	X	Check
Interest	$ 15,000	$14,000	$12,000	$255,000
Salary	150,000	--	--	98,000
Ratio	90,000	84,000	72,000	84,000
Total	$255,000	$98,000	$84,000	$437,000

3. Each should receive $\frac{1}{2}$ of the $64,000 = $32,000 each

5. 2 + 3 + 1 = 6 shares

(a) $Q = \frac{2}{6}; \quad \frac{1}{3} \times \$42,000 = \$14,000$

   $R = \frac{3}{6}; \quad \frac{1}{2} \times \$42,000 = \$21,000$

   $S = \frac{1}{6}; \quad \frac{1}{6} \times \$42,000 = \frac{\$\ 7,000}{\$42,000}$

(b) $Q = \frac{1}{3} \times (\$36,000) = (\$12,000)$

   $R = \frac{1}{2} \times (\$36,000) = (\$18,000)$

   $S = \frac{1}{6} \times (\$36,000) = \frac{(\$\ 6,000)}{(\$36,000)}$

7. Total salaries = $40,000 × 3 partners = $120,000
   $200,000 profit − $120,000 salaries = $80,000 to be divided
   25%/30%/45%

   Bob   = 25% × $80,000 = $20,000

   Chris = 30% × $80,000 = $24,000

   Doug = 45% × $80,000 = $36,000
                           $80,000

SUMMARY

	Bob	Chris	Doug	Check
Salary	$40,000	$40,000	$40,000	$60,000
% Distrib.	20,000	24,000	36,000	64,000
Total	$60,000	$64,000	$76,000	76,000
				$200,000

9. Total of their *initial* investments = $16,000 + $24,000
                                         = $40,000

   Julie $= \frac{\$16,000}{\$40,000} = \frac{2}{5}; \frac{2}{5} × \$75,000 = \$30,000$

   Greta $= \frac{\$24,000}{\$40,000} = \frac{3}{5}; \frac{3}{5} × \$75,000 = \$45,000$
                                                          $75,000

11. Total salaries = $6,000 × 2 partners = $12,000
    $64,800 profit − $12,000 salaries = $52,800 to be divided
    in the ratio of second-year investments.

    Julie $= \frac{\$20,000}{\$44,000} = \frac{5}{11}; \frac{5}{11} × \$52,800 = \$24,000$

    Greta $= \frac{\$24,000}{\$44,000} = \frac{6}{11}; \frac{6}{11} × \$52,800 = \$28,800$

SUMMARY

	Julie	Greta	Check
Salary	$ 6,000	$ 6,000	$30,000
Ratio	24,000	28,800	34,800
Total	$30,000	$34,800	$64,800

13. (a) Tom's average investment:

Date	Change	Amount of Investment		Mos. Invested		
January 1		$20,000	×	5	=	$100,000
June 1	-$4,000	16,000	×	3	=	48,000
Sept. 1	+ 1,000	17,000	×	4	=	68,000
				12		$216,000

$$\frac{\$216,000}{12 \text{ mos.}} = \$18,000 \text{ Tom's average investment}$$

(b) $26,000 Steve's average investment
   +18,000 Tom's average investment
   $44,000 Total average investment

$$\text{Steve} = \frac{\$26,000}{\$44,000} = \frac{13}{22}; \quad \frac{13}{22} \times \$77,000 = \$45,500$$

$$\text{Tom} \quad = \frac{\$18,000}{\$44,000} = \frac{9}{22}; \quad \frac{9}{22} \times \$77,000 = \frac{\$31,500}{\$77,000}$$

15. (a) Pete's average investment:

Date	Change	Amount of Investment		Mos. Invested		
January 1		$25,000	×	2	=	$ 50,000
March 1	-$3,000	22,000	×	4	=	88,000
July 1	+ 6,000	28,000	×	6	=	168,000
				12		$306,000

$$\frac{\$306,000}{12 \text{ mos.}} = \$25,500 \text{ Pete's average investment}$$

(b) Interest: Mark = 6% × $22,000 = $1,320
              Pete = 6% × $25,500 = $1,530

$30,000 Total profit
- 2,850 Total interest
$27,150 To be shared equally

$$\frac{\$27,150}{2} = \$13,575 \text{ each}$$

15. (b) (Continued)

SUMMARY

	Mark	Pete	Check
Interest	$ 1,320	$ 1,530	$14,895
Ratio	13,575	13,575	15,105
Total	$14,895	$15,105	$30,000

17. (a) Interest: Margaret = 6% × $16,000 = $ 960
                   Katherine = 6% × $30,000 = $1,800

    $14,000 Total profit
    - 2,760 Total interest
    - 5,000 Salary to Margaret
    $ 6,250 To be divided equally

$$\frac{\$6,250}{2} = \$3,120 \text{ each}$$

SUMMARY

	Margaret	Katherine	Check
Interest	$ 960	$1,800	$ 9,080
Salary	5,000	--	4,920
Ratio	3,120	3,120	$14,000
Total	$9,080	$4,920	

   (b) $7,000 Total profit
        -2,760 Total interest
        -5,000 Salary to Margaret
       ($ 760) To be divided equally

$$\frac{(\$760)}{2} = (\$380)$$

SUMMARY

	Margaret	Katherine	Check
Interest	$ 960	$1,800	$5,580
Salary	5,000	--	1,420
Ratio	(380)	(380)	$7,000
Total	$5,580	$1,420	

19. (a) Interest: Tim   = 7% × $10,000 = $700
                   Chrisy = 7% × $ 9,000 = $630
                   Barbara = 7% × $ 8,000 = $560

   $27,390 Total profit
   - 1,890 Total interest
   -12,000 Total salaries
   $13,500 To be divided in the ratio of investments

   Tim   = $\frac{10}{27}$ × $13,500 = $5,000

   Chrisy = $\frac{9}{27}$ × $13,500 = $4,500

   Barbara = $\frac{8}{27}$ × $13,500 = $4,000

SUMMARY

	Tim	Chrisy	Barbara	Check
Interest	$ 700	$ 630	$ 560	$ 9,700
Salary	4,000	4,000	4,000	9,130
Ratio	5,000	4,500	4,000	8,560
Total	$9,700	$9,130	$8,560	$27,390

(b) $13,350 Total profit
     - 1,890 Total interest
     -12,000 Total salaries
     ($ 540) To be divided in the ratio of investments

   Tim   = $\frac{10}{27}$ × ($540) = ($200)

   Chrisy = $\frac{9}{27}$ × ($540) = ($180)

   Barbara = $\frac{8}{27}$ × ($540) = ($160)

SUMMARY

	Tim	Chrisy	Barbara	Check
Interest	$ 700	$ 630	$ 560	$ 4,500
Salary	4,000	4,000	4,000	4,450
Ratio	(200)	(180)	(160)	4,400
Total	$4,500	$ 4,450	$4,400	$13,350

21. (a) Interest: Richard = 5% × $20,000 = $1,000
             Sarah   = 5% × $40,000 = $2,000
             Ted   = 5% × $30,000 = $1,500

$16,000 Total profit
- 4,500 Total interest
-10,000 Total salaries
$ 1,500 To be divided in a 1:3:2 ratio

1 + 3 + 2 = 6 shares

Richard = $\frac{1}{6}$ × $1,500 = $250

Sarah   = $\frac{1}{2}$ × $1,500 = $750

Ted   = $\frac{1}{3}$ × $1,500 = $500

SUMMARY

	Richard	Sarah	Ted	Check
Interest	$1,000	$2,000	$1,500	$ 1,250
Salary	--	5,000	5,000	7,750
Ratio	250	750	500	7,000
Total	$1,250	$7,750	$7,000	$16,000

(b) $13,300 Total profit
    - 4,500 Total interest
    -10,000 Total salaries
   ($ 1,200) To be divided in a 1:3:2 ratio

1 + 3 + 2 = 6 shares

Richard = $\frac{1}{6}$ × ($1,200) = ($200)

Sarah   = $\frac{1}{2}$ × ($1,200) = ($600)

Ted   = $\frac{1}{3}$ × ($1,200) = ($400)

SUMMARY

	Richard	Sarah	Ted	Check
Interest	$1,000	$2,000	$1,500	$ 800
Salary	--	5,000	5,000	6,400
Ratio	(200)	(600)	(400)	6,100
Total	$ 800	$6,400	$6,100	$13,350

Section 1

1.

	Trade Discounts	% Paid (or Net Cost Rate Factor)	List	Net Cost	Single Equivalent Discount %
(a)	20%	80%	$ 64	$ 51.20	x
(b)	25%	75%	240	180.00	x
(c)	40%	60%	170	102.00	x
(d)	25%	75%	96	72.00	x
(e)	20%, 25%	60%	290	174.00	40%
(f)	$22\frac{2}{9}$%, 10%	$\frac{7}{10}$ or 70%	150	105.00	30%
(g)	$37\frac{1}{2}$%, 30%	$\frac{7}{16}$ or $43\frac{3}{4}$%	160	70.00	$56\frac{1}{4}$%
(h)	20/40/30	33.6%	500	168.00	66.4%

(a) Net cost rate factor = 0.80
   % Pd × L = N
   0.80 × $64 =
      $51.20 = N

(b) Net cost rate factor = 0.75 or $\frac{3}{4}$
      % Pd × L = N
   0.75 × $240 =
         $180 = N

(c) Net cost rate factor = 0.60
   % Pd × L = N
   0.60 × L = $102

   $L = \dfrac{102}{0.60}$

   L = $170

1. (Continued)

   (d) Net cost rate factor = 75%

   Trade discount = 25%

   % Pd × L = N

   $$0.75L = 72$$

   $$L = 72 \div .75$$

   $$L = 96$$

   (e) Net cost rate factor = (0.80)(0.75) = 0.60
   % Pd × L = N
   0.60 × $290 =
   $174 = N

   Single equivalent discount:
   100%
   $\underline{-60}$
   40%

   (f) Net cost rate factor = $\left(\dfrac{7}{9}\right)\left(\dfrac{9}{10}\right) = \dfrac{7}{10}$ or 0.70

   % Pd × L = N

   $\dfrac{7}{10}$ × $150 =

   $105 = N

   Single equivalent discount:
   100%
   $\underline{-70}$
   30%

   (g) Net cost rate factor = $\left(\dfrac{5}{8}\right)\left(\dfrac{7}{10}\right) = \dfrac{7}{16}$ or 0.4375

   % Pd × L = N

   $\dfrac{7}{16}L = \$70$

   $L = 70\left(\dfrac{16}{7}\right)$

   $L = \$160$

1. (g) (Continued)

    Single equivalent discount:
    100.00%
    −43.75
    56.25%

    (h) Net cost rate factor = (0.80)(0.60)(0.70) = 0.336
    % Pd × L = N
    0.336 L = $168
    L = $500

    Single equivalent discount:
    100.0%
    −33.6
    66.4%

3. (a)   % Pd × L = N   (b)   % Pd × L = N
    0.55 × $439 =   0.65 × $500 =
        $241.45 = N $325 = N

    (c) % Pd × L = N   (d) % Pd × L = N

    .625 × $62.50 = $\frac{8}{9}$ × $18 =

        $39.06 = N $16 = N

5.   20 × $7   = $140.00
       36 × $3   = 108.00
        5 × $6   =   30.00
     10.5 × $8   =   84.00
       12 × $2   =   24.00
    Total list $386.00
    Less: 25%   − 96.50
    Net $289.50
    Plus: Freight   + 27.00
    Total due $316.50

7. (a) 1. Net cost rate factor = (0.8)(0.8) = 0.64

       2.   % Pd × L = N
          0.64 × $45 =
              $28.80 = N

       3. Single equivalent discount = 100% − 64% = 36%

7. (Continued)

(b) 1. Net cost rate factor = (0.85)(0.75) = 0.6375

2.　　% Pd × L = N
　0.6375 × $20 =
　　　　$12.75 = N

3. Single equivalent discount = 100% - 63.75% = 36.25%

(c) 1. Net cost rate factor = $(0.9)(0.745)\left(\dfrac{2}{3}\right)$ = 0.447

2.　　% Pd × L = N
　0.447 × $40 =
　　　　$17.88 = N

3. Single equivalent discount = 100% - 44.7% = 55.3%

(d) 1. Net cost rate factor = (0.75)(0.80)(0.85) = 0.51

2.　　% Pd × L = N
　0.51 × $200 =
　　　　$102 = N

3. Single equivalent discount = 100% - 51% = 49%

9. Cotrone Co.:
Net cost rate factor = (0.75)(0.80)(0.90) = 0.54
Single equivalent discount = 100% - 54% = 46%

Hodgkins:
Net cost rate factor = (0.85)(0.60) = 0.45
Single equivalent discount = 100% - 45% = 55%

Thus, Hodgkins Distributors offers a better discount.

11.　　% Pd × L = N
(0.8)(0.85)L = $3.41
　　0.68L = 3.41
　　　　L = $\dfrac{3.41}{0.68}$

　　　　L = $5.01

13.　　% Pd × L = N
(0.65)(0.9)L = $29.25
　　0.585L = 29.25
　　　　L = $\dfrac{29.25}{0.585}$

　　　　L = $50

15. % Pd × L = N

$$X(\$16) = \$12$$

$$X = \frac{12}{16}$$

$$X = 75\%$$

Thus, 100% - 75% = 25%

17. (a)   % Pd × L = N

$$(0.8)(X)(\$60) = \$42$$

$$48X = 42$$

$$X = \frac{42}{48}$$

$$X = 87.5\%$$

Thus, additional discount
needed = 100% - 87.5%
        = 12.5%

(b) Net cost rate factor =

$$(0.8)(0.875) = 0.70$$

Single equivalent
discount = 100% - 70%
         = 30%

19. (a) Crabtree Pet Supplies:
% Pd × L = N

$$\frac{5}{6} \times \$30 =$$

$$\$25 = N$$

Hannah's Pet Mart:
   % Pd × L = N
0.65 × $35 =
     $22.75 = N

(b)   % Pd × L = N

$$\left(\frac{5}{6}\right)(X)(\$30) = \$22.75$$

$$25X = 22.75$$

$$X = 91\%$$

So, additional discount
needed by Crabtree =
100% - 91% = 9%

21. (a) Shenandoah Furniture:
        % Pd × L = N

        $\dfrac{2}{3}$ × \$540 =

        \$360 = N

    Dickens Co.:
        % Pd × L = N
        0.8 × \$480 = N
              \$384 = N

(b)   % Pd × L = N

    $(0.8)(X)($480) = 360

              384X = 360

                X = 93.75%

    So, additional discount
    needed by Dickens =
    100% - 93.75% = 6.25%

23. (a) Thelma's:
          % Pd × L = N
        0.875 × \$200 =
              \$175 = N

    Johnson:
          % Pd × L = N
        0.50 × \$280 =
              \$140 = N

(b)   % Pd × L = N
    $(0.875)(X)($200) = 140
              175X = 140
                X = 80%

    So, additional discount
    needed by Thelma's
    Accessories = 100% - 80%
                = 20%

Section 2

1. (a) 2%
   (b) 3%
   (c) 1%
   (d) 2%
   (e) No discount
   (f) 2%

3. (a) No discount
       Amount due = \$824

(b)   % Pd × L = N
         (0.98)\$600 =
               \$588 = N

   (c)   % Pd × L = N
           (0.98)\$325 =
               \$318.50 = N

(d)   %Pd × L = N
           (0.99)\$780 =
               \$772.20 = N

   (e)   % Pd × L = N
           (0.97)\$450 =
               \$436.50 = N

(f)   % Pd × L = N
           (0.97)\$860 =
               \$834.20 = N

3. (Continued)

    (g)   % Pd × L = N
        (0.96)$550 =
            $528 = N

5.

	Amount of Invoice	Sales Terms	Credit Toward Account	Net Payment Made	Amount Still Due
(a)	$2,000	2/10, n/30	$1,500	$1,470.00	$ 500
(b)	860	1/10, n/30	300	297.00	560
(c)	670	2/10, n/60	500	490.00	170
(d)	790	3/15, n/30	400	388.00	390
(e)	560	4/10, n/60	360	345.60	200

(a) % Pd × Cr = N   (b)   % Pd × Cr = N
    (0.98)$1,500 =   (0.99)$300 =
        $1,470 = N $297 = N
  Amount due = $2,000 - $1,500 Amount due = $860 - $300
           = $500   = $560

(c) % Pd × Cr = N   (d) % Pd × Cr = N
    (0.98)Cr = $490   (0.97)Cr = $388
        Cr = $500   Cr = $400
  Amount due = $670 - $500   Amount due = $790 - $400
           = $170   = $390

(e) Credit to account = $560 - $200 = $360
     % Pd × L = N
   (0.96)$360 =
     $345.60 = N

7. Total cost $383.20   % Pd × L = N
  Less: Freight   - 33.20 (0.8)(0.8)(0.97)$350 =
  Merchandise cost $350.00 $217.28 = N

  Cost after discount $217.28
  Add back freight   + 33.20
  Total amount due $250.48

9. 8 × $ 8.00 = $ 64   % Pd × L = N
   10 × 18.00 = 180   (0.7)(0.75)(0.98)$334 =
   30 ×   1.20 =   36 $171.84 = N
   36 ×   1.50 =   54
                 $334

    Cost after discounts $171.84
    Add: freight   + 12.95
    Total amount due $184.79

11.   Discount   Net Amount Due
    March 3 $ 850   No discount $ 850.00
    March 10   1,000   1% × $1,000   990.00
    March 15   620   2% × $ 620   607.60
                         Total due $2,447.60

13. (a) $\frac{1}{4}$ of $1,200 = $300 credit desired

         % Pd × Cr = N
       (0.98)$300 =
             $294 = N

    (b) Balance due = $1,200 - $300 = $900

          % Pd × L = N
        (0.99)$900 =
              $891 = N

    (c) Total paid = $294 + $891 = $1,185

15. (a) % Pd × Cr = N   (b) Balance due:
        (0.97)Cr = $421.95 $870
              Cr = $435   -435
                                     $435

Section 1

1. (a) 1. C + M = S
    $80 + M = $95
      M = $15

   2. OH = 10% × C
      = (0.10)$80
    OH = $8

  3. M = OH + P
   $15 = $8 + P
    $7 = P

(b) 1. C + M = S
    $55 + M = $75
      M = $20

   2. OH = 40% × C
      = (0.4)$55
    OH = $22

  3. M = OH + P
   $20 = $22 + P
   -$2 = P (loss)

(c) 1. C + M = S
    $60 + M = $89
      M = $29

   2. OH = 20% × C
      = (0.2)$60
    OH = $12

  3. M = OH + P
   $29 = $12 + P
   $17 = P

(d) 1. C + M = S
    $33 + M = $40
      M = $7

   2. OH = 25% × S
      = (0.25)$40
    OH = $10

  3. M = OH + P
   $7 = $10 + P
   -$3 = P (loss)

(e) 1. C + M = S
    $72 + M = $96
      M = $24

   2. OH = 15% × S
      = (0.15)$96
    OH = $14.40

  3. M = OH + P
   $24 = $14.40 + P
  $9.60 = P

3.

	% Markup on Cost	Spf	Cost	Selling Price	Markup	% Markup on Selling Price
(a)	30 %	<u>1.3</u>	$90	$<u>117</u>	$<u>27</u>	$28\frac{4}{7}$ %
(b)	20	<u>1.2</u>	60	<u>72</u>	<u>12</u>	$16\frac{2}{3}$
(c)	$12\frac{1}{2}$	$\frac{9}{8}$ or 1.125	32	<u>36</u>	<u>4</u>	$11\frac{1}{9}$
(d)	25	<u>1.25</u>	<u>12</u>	15	<u>3</u>	<u>20</u>
(e)	60	<u>1.6</u>	<u>40</u>	64	<u>24</u>	$37\frac{1}{2}$
(f)	$22\frac{2}{9}$	$\frac{11}{9}$ or $1.22\frac{2}{9}$	54	66	<u>12</u>	<u>18.18</u>
(g)	$33\frac{1}{3}$	$\frac{4}{3}$ or $1.33\frac{1}{3}$	<u>27</u>	36	9	<u>25</u>

(a) Spf = 1 + 0.30 = 1.3

$$1.3 \times C = S \qquad M = S - C \qquad \underline{\quad}\% \text{ of } S = M$$
$$1.3(\$90) = \qquad = \$117 - \$90 \qquad \underline{\quad}\%(\$117) = \$27$$
$$\$117 = S \qquad M = \$27 \qquad 117x = 27$$
$$x = 23.08\%$$

(b) Spf = 1 + .20 = 1.20

$$\text{Spf} \times C = S \qquad M = S - C \qquad \underline{\quad}\% \text{ of } S = M$$
$$1.2(\$60) = \qquad = \$72 - \$60 \qquad \underline{\quad}\%(\$72) = \$12$$
$$\$72 = S \qquad M = \$12 \qquad 72x = 12$$
$$x = \frac{1}{6} \text{ or}$$
$$16\frac{2}{3}\%$$

(c) Spf = $1 + \frac{1}{8} = \frac{9}{8}$

$$\text{Spf} \times C = S \qquad M = S - C \qquad \underline{\quad}\% \text{ of } S = M$$
$$\qquad\qquad = \$36 - \$32 \qquad \underline{\quad}\%(\$36) = \$4$$
$$\frac{9}{8}(\$32) = \qquad M = \$4 \qquad 36x = 4$$
$$\$36 = S \qquad\qquad x = \frac{1}{9} \text{ or}$$
$$11\frac{1}{9}\%$$

3. (Continued)

(d) Spf = 1 + .25 = 1.25

$$\begin{array}{lll}
\text{Spf} \times \text{C} = \text{S} & \text{M} = \text{S} - \text{C} & \underline{\quad}\% \text{ of S} = \text{M} \\
1.25\text{C} = \$15 & \quad = \$15 - \$12 & \underline{\quad}\%(\$15) = \$3 \\
\text{C} = \$12 & \text{M} = \$3 & 15x = 3
\end{array}$$

$$x = \frac{1}{5} \text{ or}$$

$$20\%$$

(e) Spf = 1 + .60 = 1.60

$$\begin{array}{lll}
\text{Spf} \times \text{C} = \text{S} & \text{M} = \text{S} - \text{C} & \underline{\quad}\% \text{ of S} = \text{M} \\
1.6\text{C} = \$64 & \quad = \$64 - \$40 & \underline{\quad}\%(\$64) = \$24 \\
\text{C} = \$40 & \text{M} = \$24 & 64x = 24 \\
& & x = 37.5\%
\end{array}$$

(f)
$$\begin{array}{lll}
\text{M} = \text{S} - \text{C} & \underline{\quad}\% \text{ of C} = \text{M} & \underline{\quad}\% \text{ of S} = \text{M} \\
\text{M} = \$66 - \$54 & \underline{\quad}\%(\$54) = \$12 & \underline{\quad}\%(\$66) = \$12 \\
\text{M} = \$12 & 54x = 12 & 66x = 12 \\
& & x = 18.18\%
\end{array}$$

$$x = \frac{2}{9} \text{ or}$$

$$22\frac{2}{9}\%$$

$$\text{Spf} = 1 + \frac{2}{9} = \frac{11}{9} \quad \text{or} \quad 1.22\frac{2}{9}$$

(g)
$$\begin{array}{lll}
\text{C} = \text{S} - \text{M} & \underline{\quad}\% \text{ of C} = \text{M} & \underline{\quad}\% \text{ of S} = \text{M} \\
\quad = \$36 - \$9 & \underline{\quad}\%(\$27) = \$9 & \underline{\quad}\%(\$36) = \$9 \\
\text{C} = \$27 & 27x = 9 & 36x = 9
\end{array}$$

$$x = \frac{1}{3} \text{ or} \qquad x = \frac{1}{4} \text{ or}$$

$$33\frac{1}{3}\% \qquad\qquad 25\%$$

$$\text{Spf} = 1 + \frac{1}{3} = \frac{4}{3} \quad \text{or} \quad 1.33\overline{3}$$

5. (a) Spf = 1.6   (b)   Spf × C = S   (c) M = S - C
                          1.6($90) =   = $144 - $90
                              $144 = S   M = $54

7. Cost per pair: $\frac{\$48}{12}$ = $4   M = S - C
                                                  = $5 - $4
                                                M = $1

   (a) ___% of C = M   (b) ___% of S = M
         ___%($4) = $1   ___%($5) = $1
              4x = 1   5x = 1
               x = $\frac{1}{4}$ x = $\frac{1}{5}$
               x = 25%   x = 20%

9. (a) Spf = 1 + .37$\frac{1}{2}$ = 1.375   (b)   Spf × C = S

       (or $\frac{11}{8}$) $\frac{11}{8}$($8)   = $11

                                                 $\frac{11}{8}$($4.80)   = $ 6.60

                                                 $\frac{11}{8}$($6.40)   = $ 8.80

                                                 $\frac{11}{8}$($16.40) = $22.55

   (c) C + M = S   ___% of S = M
       $8 + M = $11   ___%($11) = $3
            M = $3   11x = 3
                                                 x = 27$\frac{3}{11}$%

11. (a)   C + M = S   (b)   C + M = S
         C + 30%C =   C + 25%C =
              1.3C = $117   1.25C = $150
                 C = $90   C = $120

     M = S - C   ___% of S = M
       = $150 - $120   ___%($150) = $30
     M = $30   150x = 30
                                               x = 20%

-145-

13. (a)   % Pd × L = Net cost   % Pd × L = Net selling price
        0.50($70) = N   0.75($70) = N
             $35 = N $52.50 = N

    (b)   C + M = S   ___% of C = M
        $35 + M = $52.50   ___%($35) = $17.50
              M = $17.50   35x = 17.50
                                            x = 50%

    (c)   ___% of S = M
        ___%($52.50) = $17.50
             52.50x = 17.50

                  x = $33\frac{1}{3}$%

15. Total cost = $345 + $15 = $360

    Cost per detector = $\frac{\$360}{15}$ = $24

    (a)   C + M = S   (b) M = S - C   ___% of S = M
         C + 25%C =   = $30 - $24   ___%($30) = $6
        1.25($24) =   M = $6   30x = 6
             $30 = S   x = 20%

17. Total cost = $120 + $24 = $144

    Cost per bottle = $\frac{\$144}{6}$ = $24

    (a)   C + M = S   (b) M = S - C   ___% of S = M
         C + 60%C =   = $38.40 - $24   ___%($38.40) =
                           $14.40
        1.6($24) =   M = $14.40   38.40x =
                                                 14.40
        $38.40 = S
                                                 x =
                                            $37\frac{1}{2}$%

Section 2

1.

	% Markup on Selling Price	Spf	Cost	Selling Price	Markup	% Markup on Cost
(a)	25 %	$\dfrac{4}{3}$	$ 90	$ 120	$30	$33\frac{1}{3}$%
(b)	$37\frac{1}{2}$	$\dfrac{8}{5}$ or 1.6	35	56	21	60
(c)	$33\frac{1}{3}$	$\dfrac{3}{2}$ or 1.5	1,000	1,500	500	50
(d)	50	$\dfrac{1}{.5}$ or 2	6	12	6	100
(e)	10	$\dfrac{10}{9}$	18	20	2	$11\frac{1}{9}$

(a) $Spf = \dfrac{4}{3}$

$C + M = S$
$\$90 + M = \$120$
$M = \$30$

____% of $C = M$
____%($90) = $30
$90x = 30$
$x = 33\frac{1}{3}$%

$Spf \times C = S$

$\dfrac{4}{3}(\$90) =$

$\$120 = S$

(b) $Spf = \dfrac{8}{5}$ or 1.6

$C + M = S$
$\$35 + M = \$56$
$M = \$21$

____% of $C = M$
____%($35) = $21
$35x = 21$
$x = 60$%

$Spf \times C = S$

$\dfrac{8}{5}(\$35) =$

$\$56 = S$

1. (Continued)

(c) $Spf = \dfrac{3}{2}$ or 1.5 $\quad$ C + M = S $\qquad$ ___% of C = M

$\qquad\qquad\qquad\qquad\quad$ \$1,000 + M = \$1,500 $\quad$ ___%(\$1,000) = \$500

$\qquad\qquad\qquad\qquad\qquad\qquad\quad$ M = \$500 $\qquad\qquad$ 1,000x = 500

$\qquad$ Spf × C = S $\qquad\qquad\qquad\qquad\qquad\qquad\qquad$ x = 50%

$\qquad\quad \dfrac{3}{2}C = \$1,500$

$\qquad\qquad\;\; C = \$1,000$

(d) $Spf = \dfrac{1}{.5}$ or 2 $\qquad$ C + M = S $\qquad$ ___% of C = M

$\qquad\qquad\qquad\qquad\qquad$ \$6 + M = \$12 $\qquad$ ___%(\$6) = \$6

$\qquad\qquad\qquad\qquad\qquad\qquad\quad$ M = \$6 $\qquad\qquad\quad$ 6x = 6

$\qquad$ Spf × C = S $\qquad\qquad\qquad\qquad\qquad\qquad\quad$ x = 100%

$\qquad\quad \dfrac{1}{.5}C = \$12$

$\qquad\qquad\; C = \$6$

(e) $\quad$ C + M = S $\qquad$ ___% of S = M $\qquad$ ___% of C = M

$\quad$ \$18 + \$2 = $\qquad$ ___%(\$20) = \$2 $\qquad$ ___%(\$18) = \$2

$\qquad\qquad$ \$20 = S $\qquad\qquad\quad$ 20x = 2 $\qquad\qquad\quad$ 18x = 2

$\qquad\qquad\qquad\qquad\qquad\qquad\quad$ x = 10% $\qquad\qquad\qquad$ $x = 11\dfrac{1}{9}$%

$\quad Spf = \dfrac{10}{9}$

3. (a) $Spf = 1.42857$ or $\dfrac{10}{7}$

(b) $\quad$ Spf × C = S $\qquad\qquad$ (c) M = S − C

$\quad \dfrac{10}{7}(\$280) =$ $\qquad\qquad\qquad\qquad$ = \$400 − \$280

$\qquad\qquad\qquad\qquad\qquad\qquad\qquad$ M = \$120

$\qquad$ \$400 = S

5. (a) ___% of S = M $\qquad\qquad$ (b) $\quad$ C + M = S

$\qquad 37\dfrac{1}{2}$%S = \$15 $\qquad\qquad\qquad$ C + \$15 = \$40

$\qquad\quad$ 0.375S = 15 $\qquad\qquad\qquad\qquad\quad$ C = \$25

$\qquad\qquad\qquad$ S = \$40

7. (a) Spf = 2.5 or $\frac{5}{2}$

    (b) $\quad$ Spf × C = S
    $$2.5(\$5) \quad = \$12.50$$
    $$2.5(\$6.40) = \$16.00$$
    $$2.5(\$7.20) = \$18.00$$
    $$2.5(\$8.50) = \$21.25$$

    (c) M = S − C
    $$= \$12.50 - \$5.00$$
    $$M = \$7.50$$

    $\underline{\quad}$% of C = M
    $$\underline{\quad}\%(\$5) = \$7.50$$
    $$5x = 7.50$$
    $$x = 150\%$$

9. (a) Total sales revenue:
    $$20 \ @ \ \$60 = \$1,200$$
    $$16 \ @ \ \$54 = \quad 864$$
    $$9 \ @ \ \$46 = \quad 414$$
    $$5 \ @ \ \$36 = \quad \underline{\quad 180}$$
    $$\text{Total} \quad\quad \$2,658$$

    Total sales $\quad$ \$2,658
    Total cost $\quad$ $\underline{-2,215}$
    Total margin $\quad$ \$ 443

    (b) $\underline{\quad}$% of C = M
    $$\underline{\quad}\%(\$2,215) = \$443$$
    $$2,215x = 443$$
    $$x = 20\%$$

    (c) $\underline{\quad}$% of S = M
    $$\underline{\quad}\%(\$2,658) = \$443$$
    $$2,658x = 443$$
    $$x = 16\frac{2}{3}\%$$

11. (a) 50% on cost:

    Spf = 1 + .50 = 1.50

    50% on selling price:

    Spf = $\frac{1}{.5}$ = 2.0

    Larger Spf belongs to 50% on selling price, so it will yield the larger gross profit.

    (b) 20% on cost:

    Spf = 1 + .20 = 1.20

    28% on selling price:

    Spf = $\frac{1}{.72}$ = $1.3\overline{8}$

    28% on selling price yields a larger Spf, thus a larger gross profit.

13. (a) $M = 37\frac{1}{2}\%S = \frac{3}{8}S$ (b) Spf × C = S

$$Spf = \frac{8}{5} \text{ or } 1.6$$

(c)   C + M = S
$$\$4,590 + M = \$7,344$$
$$M = \$2,754$$

$\qquad$ 1.6C = \$7,344
$\qquad$ C = \$4,590

___% of C = M
___%(\$4,590) = \$2,754
$\qquad$ 4,590x = 2,754
$\qquad\qquad$ x = 60%

15. (a)   C + M = S
$$C + 35\%S = S$$
$$C = 0.65(\$340)$$
$$C = \$221$$

(b) $Spf = \dfrac{1}{.65}$ or $\dfrac{20}{13}$ or 1.538

17. % Pd × L = Net cost
$$\frac{5}{6}(\$42) = N$$
$$\$35 = N$$

M = OH + P
M = 25%S + 5%S
M = 30%S

C + M = S
C + 30%S = S
$\qquad$ C = 70%S
$\qquad$ \$35 = 0.7S
$\qquad$ \$50 = S

Section 3

1.	Quantity Bought	Cost per Unit	Total Cost	Markup on Cost	Total Sales	Pct. to Spoil	Amt. to Sell	Selling Price
(a)	70 lbs.	\$0.60	\$ 42	20 %	\$ 50.40	10%	63 lbs.	\$0.80 lb.
(b)	20 dz.	1.40	28	25	35.00	5	19 dz.	1.85 dz.
(c)	50	4.00	200	$12\frac{1}{2}$	225.00	4	48	4.69

(a) Total cost = 70 × \$0.60 = \$42
Amount to spoil = 10% × 70 lbs. = 7 lbs.
Amount to sell = 70 lbs. - 7 lbs. = 63 lbs.

$\qquad$ C + M = S
$\qquad$ C + 20%C =
$\qquad$ 1.2(\$42) =
$\qquad$ \$50.40 = S

No. to sell × price = Total sales
$\qquad$ 63p = \$50.40
$\qquad$ p = \$ 0.80

1. (Continued)

(b) Total cost = 20 dz. × $1.40 = $28
    Amount to spoil = 5% × 20 dz. = 1 dz.
    Amount to sell = 20 dz. - 1 dz. = 19 dz.

$$C + M = S \qquad \text{No. to sell} \times \text{price} = \text{Total sales}$$
$$C + 25\%C = \qquad\qquad\qquad 19p = \$35$$
$$1.25(\$28) = \qquad\qquad\qquad\quad p = \$1.85$$
$$\$35 = S$$

(c) Total cost = 50 × $4 = $200
    Amount to spoil = 4% × 50 = 2
    Amount to sell = 50 - 2 = 48

$$C + M = S \qquad \text{No. to sell} \times \text{price} = \text{Total sales}$$
$$\qquad\qquad\qquad\qquad\qquad 48p = \$225$$
$$C + \frac{1}{8}C = \qquad\qquad\qquad\quad p = \$4.69$$
$$\frac{9}{8}(\$200) =$$
$$\$225 = S$$

3.

	Quantity Bought	Cost per Unit	Total Cost	Markup on Cost	Total Sales (Cont. below)
(a)	80 lbs.	$5.00	$400	20%	$480.00
(b)	25	3.00	75	40	105.00

	Amount at Reg. Price	% at Reduced Price	Amount at Reduced Price	Reduced Price	Reg. S.P.
(a)	76 lbs.	5%	4 lbs.	$4.00	$6.11/lb.
(b)	23	8	2	2.50	4.35

(a) Total cost = 80 × $5.00 = $400
    5% × 80 lbs. = 4 lbs. sold at reduced price
    80 lbs. - 4 lbs. = 76 lbs. to sell at regular price

3. (a) (Continued)

$$\begin{array}{ll} C + M = S \\ C + 20\%C = \\ 1.2(\$400) = \\ \$480 = S \end{array} \qquad \begin{array}{l} \text{Regular + Reduced = Total sales} \\ 76p + 4(\$4) = \$480 \\ 76p + 16 = 480 \\ 76p = 464 \\ p = \$6.11 \end{array}$$

(b) Total cost = 25 × \$3 = \$75
8% × 25 = 2 to sell at reduced price
25 - 2 = 23 to sell at regular price

$$\begin{array}{ll} C + M = S \\ C + 40\%C = \\ 1.4(\$75) = \\ \$105 = S \end{array} \qquad \begin{array}{l} \text{Regular + Reduced = Total sales} \\ 23p + 2(\$2.50) = \$105 \\ 23p + 5 = 105 \\ 23p = 100 \\ p = \$4.35 \end{array}$$

5. Referring to Problem 1(a):
Total sales = 70 lbs. × \$0.80 = \$56

(a)
$$\begin{array}{l} C + M = S \\ \$42 + M = \$56 \\ M = \$14 \end{array}$$

(b)
$$\begin{array}{l} \underline{\quad}\% \text{ of } C = M \\ \underline{\quad}\%(\$42) = \$14 \\ 42x = 14 \\ x = 33\frac{1}{3}\% \end{array}$$

(c)
$$\begin{array}{l} \underline{\quad}\% \text{ of } S = M \\ \underline{\quad}\%(\$56) = \$14 \\ 56x = 14 \\ x = 25\% \end{array}$$

7. Total cost = 200 lbs. × \$0.60 = \$120
Amount expected to spoil = 10% × 200 lbs. = 20 lbs.
Amount expected to sell = 200 lbs. - 20 lbs. = 180 lbs.

$$\begin{array}{ll} C + M = S \\ C + 15\%C = \\ 1.15(\$120) = \\ \$138 = S \end{array} \qquad \begin{array}{l} \text{Amount to sell × price = Total sales} \\ 180p = \$138 \\ p = \$0.77/\text{lb.} \end{array}$$

9. Total cost = 300 lbs. × $0.72 = $216
   Amount expected to spoil = 7% × 300 lbs. = 21 lbs.
   Amount expected to sell = 300 lbs. - 21 lbs. = 279 lbs.

   $$C + M = S$$
   $$C + 25\%C =$$
   $$1.25(\$216) =$$
   $$\$270 = S$$

   Amount to sell × price = Total sales
   $$279p = \$270$$
   $$p = \$0.97/lb.$$

11. Total cost = 20 doz. × $6 = $120
    Amount expected to spoil = 5% × 20 dz. = 1 doz.
    Amount expected to sell = 20 doz. - 1 doz. = 19 doz.

    $$C + M = S$$
    $$C + 45\%C =$$
    $$1.45(\$120) =$$
    $$\$174 = S$$

    No. to sell × price = Total sales
    $$19p = \$174$$
    $$p = \$9.16/doz.$$

13. Total cost = 50 loaves × $0.55 = $27.50
    10% × 50 loaves = 5 loaves will be sold at $0.40/ea.
    50 - 5 = 45 loaves to sell at regular price

    $$C + M = S$$
    $$C + 60\%C =$$
    $$1.6(\$27.50) =$$
    $$\$44 = S$$

    Regular + Reduced = Total sales
    $$45p + 5(\$0.40) = \$44$$
    $$45p + 2 = 44$$
    $$45p = 42$$
    $$p = \$0.94/ea.$$

15. Total cost = 75 × $120 = $9,000
    12% × 75 = 9 jackets will be sold at $118 each
    75 - 9 = 66 jackets left to sell at regular price

    $$C + M = S$$
    $$C + 35\%C =$$
    $$1.35(\$9,000) =$$
    $$\$12,150 = S$$

    Regular + Reduced = Total sales
    $$66p + 9(\$118) = \$12,150$$
    $$66p + 1,062 = 12,150$$
    $$66p = 11,088$$
    $$p = \$168$$

Section 1

1.

	Cost	% Regular Markup (M_1)	Reg. Price (S_1)	% Discount	Sale Price (S_2)	% Sale Markup (M_2)
(a)	$ 50	40%C	$ 70	x	$ 62.50	20%S
(b)	48	$33\frac{1}{3}$%C	64	x	57.60	$16\frac{2}{3}$S%
(c)	28	30%S	40	20 %	32.00	$12\frac{1}{2}$%S
(d)	30	25%S	40	10	36.00	$16\frac{2}{3}$%S
(e)	27	46%S	50	10	45.00	40%S
(f)	148.75	40.5%S	250	30	175.00	15%S
(g)	240	60%S	600	$33\frac{1}{3}$/20	320.00	25%S
(h)	128	60%S	320	$37\frac{1}{2}$/20	160.00	20%S

(a)

$$C + M_1 = S_1$$
$$C + 40\%C = S_1$$
$$1.4C = \$70$$
$$C = \$50$$

$$C + M_2 = S_2$$
$$C + 20\%S_2 = S_2$$
$$C = 80\%S_2$$
$$\$50 = 0.80S_2$$
$$\$62.50 = S_2$$

(b)

$$C + M_1 = S_1$$
$$C + \frac{1}{3}C = S_1$$
$$\frac{4}{3}C = \$64$$
$$C = \$48$$

$$C + M_2 = S_2$$
$$C + \frac{1}{6}S_2 = S_2$$
$$C = \frac{5}{6}S_2$$
$$\$48 = \frac{5}{6}S_2$$
$$\frac{6}{5}(48) = S_2$$
$$\$57.60 = S_2$$

1. (Continued)

(c)
$$C + M_1 = S_1$$
$$C + 30\%S_1 = S_1$$
$$C = 70\%S_1$$
$$\frac{1}{.7}(\$28) = S_1$$
$$\$40 = S_1$$

$$\% \; Pd \times L = N$$
$$80\%(\$40) = S_2$$
$$\$32 = S_2$$

$$C + M_2 = S_2$$
$$\$28 + M_2 = \$32$$
$$M_2 = \$4$$

$$\underline{\quad}\% \text{ of } S_2 = M_2$$
$$\underline{\quad}\%(\$32) = \$4$$
$$32x = 4$$
$$x = 12\frac{1}{2}\%$$

(d)
$$C + M_1 = S_1$$
$$C + 25\%S_1 = S_1$$
$$C = 75\%S_1$$
$$\$30 = 0.75S_1$$
$$\$40 = S_1$$

$$\% \; Pd \times L = N$$
$$90\%(\$40) = S_2$$
$$\$36 = S_2$$

$$C + M_2 = S_2$$
$$\$30 + M_2 = \$36$$
$$M_2 = \$6$$

$$\underline{\quad}\% \text{ of } S_2 = M_2$$
$$\underline{\quad}\%(\$36) = \$6$$
$$36x = 6$$
$$x = 16\frac{2}{3}\%$$

(e)
$$C + M_2 = S_2$$
$$C + 40\%S_2 = S_2$$
$$C = 60\%(\$45)$$
$$C = \$27$$

$$\% \; Pd \times L = N$$
$$90\%S_1 = S_2$$
$$90\%S_1 = \$45$$
$$S_1 = \$50$$

$$C + M_1 = S_1$$
$$\$27 + M_1 = \$50$$
$$M_1 = \$23$$

$$\underline{\quad}\% \text{ of } S_1 = M_1$$
$$\underline{\quad}\%(\$50) = \$23$$
$$50x = 23$$
$$x = 46\%$$

1. (Continued)

(f)
$$C + M_2 = S_2$$
$$C + 15\%S_2 = S_2$$
$$C = 85\%(\$175)$$
$$C = \$148.75$$

$$\%\text{ Pd} \times L = N$$
$$70\%S_1 = S_2$$
$$70\%S_1 = \$175$$
$$S_1 = \$250$$

$$C + M_1 = S_1$$
$$\$148.75 + M_1 = \$250.00$$
$$M_1 = \$101.25$$

$$\underline{\quad}\%\text{ of }S_1 = M_1$$
$$\underline{\quad}\%(\$250) = \$101.25$$
$$250x = 101.25$$
$$x = 40.5\%$$

(g)
$$\%\text{ Pd} \times L = \text{Net Cost}$$
$$60\%(\$400) = C$$
$$\$240 = C$$

$$C + M_2 = S_2$$
$$C + 25\%S_2 = S_2$$
$$C = 75\%S_2$$
$$\frac{1}{.75}(\$240) = S_2$$
$$\$320 = S_2$$

$$\%\text{ Pd} \times L = N$$
$$\left(\frac{2}{3}\right)\left(\frac{4}{5}\right)S_1 = S_2$$
$$\frac{8}{15}S_1 = \$320$$
$$S_1 = \$600$$

$$C + M_1 = S_1$$
$$\$240 + M_1 = \$600$$
$$M_1 = \$360$$

$$\underline{\quad}\%\text{ of }S_1 = M_1$$
$$\underline{\quad}\%(\$600) = \$360$$
$$600x = 360$$
$$x = 60\%$$

(h)
$$\%\text{ Pd} \times L = \text{Net Cost}$$
$$(80\%)(80\%)\$200 = C$$
$$0.64(200) = C$$
$$\$128 = C$$

$$C + M_2 = S_2$$
$$C + 20\%S_2 = S_2$$
$$C = 80\%S_2$$
$$\frac{1}{.8}(\$128) = S_2$$
$$\$160 = S_2$$

1. (h) (Continued)

$$\% \text{ Pd} \times L = N$$

$$\left(\frac{5}{8}\right)\left(\frac{4}{5}\right)S_1 = S_2$$

$$\frac{1}{2}S_1 = \$160$$

$$S_1 = \$320$$

$$\_\_\_\% \text{ of } S_1 = M_1$$

$$\_\_\_\%(\$320) = \$192$$

$$320x = 192$$

$$x = 60\%$$

$$C + M_1 = S_1$$

$$\$128 + M_1 = \$320$$

$$M_1 = \$192$$

3. (a)

$$C + M_1 = S_1$$

$$C + 45\%C = \$290$$

$$1.45C = 290$$

$$C = \$200$$

(b)

$$C + M_2 = S_2$$

$$C + 25\%S_2 = S_2$$

$$C = 75\%S_2$$

$$\frac{1}{.75}(\$200) = S_2$$

$$\$266.67 = S_2$$

5. (a)

$$C + M_1 = S_1$$

$$C + 37.5\%S_1 = S_1$$

$$C = 62.5\%S_1$$

$$\$800 = 0.625S_1$$

$$\$1,280 = S_1$$

(c)

$$C + M_2 = S_2$$

$$\$800 + M_2 = \$960$$

$$M_2 = \$160$$

(b)

$$\% \text{ Pd} \times L = N$$

$$75\%(\$1,280) = S_2$$

$$\$960 = S_2$$

$$\_\_\_\% \text{ of } S_2 = M_2$$

$$\_\_\_\%(\$960) = \$160$$

$$960x = 160$$

$$x = 16\frac{2}{3}\%$$

7. (a) $\quad$ C + M_1 = S_1

$\quad$ C + 40%S_1 = S_1

$\qquad\qquad$ C = 60%S_1

$\qquad\quad$ \$18 = $0.60S_1$

$\qquad\quad$ \$30 = S_1

(b) $\quad$ % Pd × L = N

$\qquad\qquad$ 75%S_1 = S_2

$\qquad$ 0.75(\$30) = S_2

$\qquad\qquad$ \$22.50 = S_2

(c) $\quad$ C + M_2 = S_2

$\quad$ \$18 + M_2 = \$22.50

$\qquad\qquad M_2$ = \$4.50

$\qquad$ ___% of S_2 = M_2

$\qquad$ ___%(\$22.50) = \$4.50

$\qquad\qquad$ 22.5x = 4.50

$\qquad\qquad$ x = 20%

9. (a) $\quad$ % Pd × L = N

$\qquad\qquad$ 75%S_1 = S_2

$\qquad$ 0.75S_1 = \$90

$\qquad\qquad S_1$ = \$120

(b) $\quad$ C + M_2 = S_2

$\quad$ C + 40%S_2 = S_2

$\qquad\qquad$ C = 60%S_2

$\qquad\qquad$ C = 0.60(\$90)

$\qquad\qquad$ C = \$54

(c) $\quad$ C + M_1 = S_1

$\quad$ \$54 + M_1 = \$120

$\qquad\qquad M_1$ = \$66

$\qquad$ ___% of S_1 = M_1

$\qquad$ ___%(\$120) = \$66

$\qquad\qquad$ 120x = 66

$\qquad\qquad$ x = 55%

11. (a) $\quad$ % Pd × L = N

$\qquad\qquad$ 87.5%S_1 = S_2

$\qquad$ 0.875S_1 = \$48

$\qquad\qquad S_1$ = \$54.86

(b) $\quad$ C + M_2 = S_2

$\quad$ C + 20%S_2 = S_2

$\qquad\qquad$ C = 80%S_2

$\qquad\qquad$ C = 0.80(\$48)

$\qquad\qquad$ C = \$38.40

(c) $\quad\quad$ C + M_1 = S_1

$\quad$ \$38.40 + M_1 = \$54.86

$\qquad\qquad M_1$ = \$16.46

$\qquad$ ___% of S_1 = M_1

$\qquad$ ___%(\$54.86) = \$16.46

$\qquad\qquad$ 54.86x = 16.46

$\qquad\qquad$ x = 30%

13. (a) $\quad$ % Pd × L = Net Cost

$\qquad$ (70%)(75%)$140 = C

$\qquad$ 0.525(140) = C

$\qquad$ $73.50 = C

(b) $\qquad$ C + M$_2$ = S$_2$

$$C + \frac{1}{3}S_2 = S_2$$

$$C = \frac{2}{3}S_2$$

$$\frac{3}{2}(\$73.50) = S_2$$

$$\$110.25 = S_2$$

(c) % Pd × L = N

$\qquad$ 60%S$_1$ = S$_2$

$\qquad$ 0.60S$_1$ = $110.25

$\qquad$ S$_1$ = $183.75

15. (a) % Pd × L = N

$\qquad$ 70%($50) = N

$\qquad$ $35 = N

(b) $\qquad$ C + M$_2$ = S$_2$

$\qquad$ C + 60%S$_2$ = S$_2$

$\qquad$ C = 40%S$_2$

$\qquad$ $35 = 0.40S$_2$

$\qquad$ $87.50 = S$_2$

(c) % Pd × L = N

$\qquad$ 87.5%S$_1$ = S$_2$

$\qquad$ 0.875S$_1$ = $87.50

$\qquad$ S$_1$ = $100

Section 2

1.

	Regular Selling Price (S_1)	Markdown %	Markdown Amt.	Sale Price (S_2)	Whole-sale cost	Over-head	Total Handling Cost	Operating Profit or (Loss)
(a)	$60	40 %	$24	$36	$22	$4	$26	$10
(b)	30	30	9	21	19	3	22	(1)
(c)	45	$22\frac{2}{9}$	10	35	23	8	31	4
(d)	18	$11\frac{1}{9}$	2	16	12	7	19	(3)
(e)	60	10	6	54	50	9	59	(5)
(f)	40	20	8	32	18	8	26	6

(a) Markdown amount = $60 × 40% = $24
   Sale price = $60 - $24 = $36
   THC = $22 + $4 = $26
   Profit = $36 - $26 = $10

(b) Markdown percent = (on right)   ___% of original = change
   Sale price = $30 - $9 = $21   ___%($30) = $9
   THC = $19 + $3 = $22 $30x$ = 9
   Loss = $22 - $21 = -$1 x = 30%

(c) Markdown amt. = $45 - $35 = $10   ___% of original = change
   Markdown percent = (on right)   ___%($45) = $10
   THC = $35 - $4 = $31 $45x$ = 10
   Wholesale cost = $31 - $8 = $23 $x = 22\frac{2}{9}$%

1. (Continued)

(d) Wholesale cost = $19 - $7 = $12   ___% of original = change
    Sale price = $19 - $3 = $16   ___%($18) = $2
    Markdown amount = $18 - $16 = $2   18x = 2
    Markdown percent = (on right) $x = 11\frac{1}{9}\%$

(e) Overhead = $59 - $50 = $9   ___% of original = change
    Sale price = $59 - $5 = $54   ___%($60) = $6
    Regular price = $54 + $6 = $60   60x = 6
    Markdown percent = (on right)   x = 10%

(f) Regular price = (on right)   % Pd × L = N
    Markdown amt. = $40 - $32 = $8   80%$S_1$ = $32
    THC = $32 - $6 = $26 S_1 = $40
    Wholesale cost = $26 - $8 = $18

3.

	Regular Selling Price(S_1)	Markdown %	Markdown Amt.	Sale Price(S_2)	Whole-sale Cost	Over-head	Total Handling Cost	Operating Loss	Gross Loss	Loss %
(a)	$36	$16\frac{2}{3}$%	$ 6	$30	$32	$3	$35	$ 5	$2	6.25%
(b)	80	40	32	48	50	6	56	8	2	4
(c)	40	25	8	32	35	8	43	11	3	8.57
(d)	90	30	27	63	70	4	74	11	7	10

(a) S_1 - MD = S_2 ___% of original = change
    $36 - $6 = S_2 ___%($36) = $6
      $30 = S_2 36x = 6
         $x = 16\frac{2}{3}\%$

3. (a) (Continued)

$$THC = C + OH$$
$$THC = \$32 + \$3$$
$$THC = \$35$$

$$THC - S_2 = OL$$
$$\$35 - \$30 = OL$$
$$\$5 = OL$$

$$GL = C - OH$$
$$GL = \$32 - \$30$$
$$GL = \$2$$

$$\underline{\quad}\% \text{ of } C = GL$$
$$\underline{\quad}\%(\$32) = \$2$$
$$32x = 2$$
$$x = 6.25\%$$

(b) $\underline{\quad}\%$ of original = change
$$40\%(\$80) = MD$$
$$\$32 = MD$$

$$S_1 - MD = S_2$$
$$\$80 - \$32 = S_2$$
$$\$48 = S_2$$

$$THC = C + OH$$
$$THC = 50 + 6$$
$$THC = 56$$

$$OL = THC - S_2$$
$$OL = \$56 - \$48$$
$$OL = \$8$$

$$GL = C - S_2$$
$$GL = \$50 - \$48$$
$$GL = \$2$$

$$\underline{\quad}\% \text{ of } C = GL$$
$$\underline{\quad}\%(\$50) = \$2$$
$$50x = 2$$
$$x = 4\%$$

(c) $\% \text{ Pd} \times L = N$
$$80\%S_1 = \$32$$
$$S_1 = \$40$$

$$S_1 - MD = S_2$$
$$\$40 - MD = \$32$$
$$MD = \$8$$

$$THC = C + OH$$
$$\$43 = C + \$8$$
$$\$35 = C$$

$$OL = THC - S_2$$
$$OL = \$43 - \$32$$
$$OL = \$11$$

$$GL = C - S_2$$
$$GL = \$35 - \$32$$
$$GL = \$3$$

$$\underline{\quad}\% \text{ of } C = GL$$
$$\underline{\quad}\%(\$35) = \$3$$
$$35x = 3$$
$$x = 8.57\%$$

3. (Continued)

(d) % Pd × L = N   MD = S_1 - S_2
        70% S_1 = \$63   MD = \$90 - \$63
             S_1 = \$90   MD = \$27

    GL = C - S_2 THC = C + OH
    \$7 = C - \$63   THC = \$70 + \$4
    \$70 = C   THC = \$74

    ___% of C = GL   OL = THC - S_2
    ___%(\$70) = \$7   OL = \$74 - \$63
          70x = 7   OL = \$11
            x = 10%

5. % Pd × L = N   THC = C + 25%C   OL = THC - S_2
        70% S_1 = S_2 THC = 125%(\$130)   OL = \$162.50 - \$139.30
    70%(\$199) = S_2 THC = \$162.50   OL = \$23.20
      \$139.30 = S_2

7. % Pd × L = N   THC = C + 20%C   OL = THC - S_2
        75% S_1 = S_2 THC = 120%(\$26)   OL = \$31.20 - \$30
      75%(\$40) = S_2 THC = \$31.20   OL = \$1.20
          \$30 = S_2

9. (a) Markdown amount = \$54 - \$45 = \$9

         ___% of original = change
             ___%(\$54) = \$9
                   54x = 9
                     x = $16\frac{2}{3}$%

9. (Continued)

(b)   % Pd × L = Net Cost

$$(80\%)(85\%)\$50 = C$$
$$\$34 = C$$

THC = C + OH   OP = S_2 − THC
THC = \$34.00 + \$5.40   OP = \$45.00 − \$39.50
THC = \$39.40   OP = \$5.60

11. (a)   C + OH + P = S_1

$$C + 20\%S + 10\%S = S_1$$

$$C = 70\%S_1$$

$$\frac{1}{.7}(\$140) = S_1$$

$$\$200 = S_1$$

(b) % Pd × L = N   OH = 20\%S_1
       $50\%S_1 = S_2$ OH = 0.20(\$200)
  $50\%(\$200) = S_2$ OH = \$40
      $\$100 = S_2$

THC = C + OH   OL = THC − S_2
THC = \$140 + \$40   OL = \$180 − \$100
THC = \$180   OL = \$80

(c) GL = C − S_2 ____% of C = GL
    GL = \$140 − \$100   ____%(\$140) = \$40
    GL = \$40   140x = 40

$$x = 28\frac{4}{7}\%$$

13. (a)

$$\% \text{ Pd} \times L = \text{Net Cost}$$
$$(80\%)(80\%)\$7.50 = C$$
$$(0.64)7.50 = C$$
$$\$4.80 = C$$

$$\frac{\$4.80}{12} = \$0.40/\text{each}$$

$$C + 30\%C + 20\%C = S_1$$
$$1.5(\$0.40) = S_1$$
$$\$0.60 = S_1$$

(b) $\text{THC} = C + OH$
$$\text{THC} = C + 30\%C$$
$$\text{THC} = 1.3(\$0.40)$$
$$\text{THC} = \$0.52$$

$$S_1 - \text{THC} = \text{Max. MD}$$
$$\$0.60 - \$0.52 = \text{MD}$$
$$\$0.08 = \text{MD}$$

(c) $\underline{\quad}\%$ of original = change
$$\underline{\quad}\%(\$0.60) = \$0.08$$
$$0.60x = 0.08$$
$$x = 13\frac{1}{3}\%$$

15. (a)

$$\% \text{ Pd} \times L = \text{Net Cost}$$
$$(60\%)(80\%)\$90 = C$$
$$\$43.20 = C$$

$$\frac{\$43.20}{12} = \$3.60/\text{each}$$

$$C + M_1 = S_1$$
$$C + 25\%C + 50\%C = S_1$$
$$175\%(\$3.60) = S_1$$
$$\$6.30 = S_1$$

(b) $\text{THC} = 125\%C$
$$\text{THC} = 125\%(\$3.60)$$
$$\text{THC} = \$4.50$$

$$S_1 - \text{THC} = \text{Max. MD}$$
$$\$6.30 - \$4.50 = \text{MD}$$
$$\$1.80 = \text{MD}$$

(c) $\underline{\quad}\%$ of original = change
$$\underline{\quad}\%(\$6.30) = \$1.80$$
$$6.30x = 1.80$$
$$x = 28.57\%$$

17. (a) Total cost = 200 × $2.80 = $560
       OH = 15% × $560 = $84
       Total handling cost = $560 + $84 = $644
       Total sales = $670 (below)

       180 × $3.50 = $630
        20 × $2.00 = + 40
                      $670

    (b) Total sales ($670) exceeds total handling cost ($644) = operating profit

    (c) OP = S - THC   ___% of C = OP
        OP = $670 - $644   ___%($560) = $26
        OP = $26   560x = 26
                                                x = 4.6%

19. (a) Total cost = 25 × $4 = $100
       Total handling cost = 130% × $100 = $130
       Total sales = $143 (below)

       18 × $6 = $108
        7 × $5 = + 35
                  $143

    (b) Total sales ($143) exceeds total handling cost ($130) = operating profit

    (c) OP = S - THC   ___% of C = OP
        OP = $143 - $130   ___%($100) = $13
        OP = $13   100x = 13
                                                x = 13%

Section 3

1. (a) Total cost before trade discount:
   $$80 \times \$20 = \$1,600$$

   Total cost after trade discount:
   $$\_\_\_\% \text{ Pd} \times L = N$$
   $$70\%(\$1,600) = N$$
   $$\$1,120 = N$$

   (b) $\dfrac{\$1,120}{2} = \$560 =$ credit for half the balance

   10 days from invoice date = 2% discount

   $$\_\_\_\% \text{ Pd} \times \text{Credit} = \text{Net payment}$$
   $$98\%(\$560) = N$$
   $$\$548.80 = N$$

   (c) 20 days from invoice date = 1% discount

   $$\_\_\_\% \text{ Pd} \times \text{Balance} = N$$
   $$99\%(\$560) = N$$
   $$\$554.40 = N$$

   (d) Total cost = $\$548.80 + \$554.40 = \$1,103.20$

   Individual cost $= \dfrac{\$1,103.20}{80} = \$13.79$ each

   (e)
   $$C + M = S$$
   $$C + 20\%C + 20\%C = S$$
   $$140\%C = S$$
   $$140\%(\$13.79) = S$$
   $$\$19.31 = S$$

1. (Continued)

   (f) ___% Pd × S_1 = S_2
      60%($19.31) = $11.59

   (g) Total revenue:
      50 × $19.31 = $ 965.50
      30 × $11.59 = +   347.70
                     $1,313.20

   (h) Total handling cost = 120%($1,103.20) = $1,323.84

       Operating loss = Total handling cost ($1,323.84) – Total
       sales ($1,313.20) = $10.64

   (i)   ___% of C = OL
      ___%($1,103.20) = $10.64
          1,103.20x =   10.64
                 x = 0.964% or 1.0%

3. (a) Total cost before trade discount:
      100 × $125 = $12,500

      Total cost after trade discount:
         ___% Pd × L = N
      (85%)(80%)$12,500 = N
              $8,500 = N

      Net cost after cash discount:
      ___% Pd × L = N
      97%($8,500) = N
          $8,245 = N

      Net merchandise cost   = $8,245
      Freight charge   = +   35
      Amount of Sept. 12 payment = $8,280

3. (Continued)

(b) Total cost \$8,280 and individual cost = $\frac{\$8,280}{100}$ = \$82.80

(c)
$$C + M = S$$
$$C + 25\%C + 15\%C = S$$
$$140\%C = S$$
$$140(\$82.80) = S$$
$$\$115.92 = S$$

(d) Reduced selling price:
$$\_\_\% \ Pd \times S_1 = S_2$$
$$50\%(\$115.92) = S_2$$
$$\$57.96 = S_2$$

(e) Total revenue:
$$75 \times \$115.92 = \$\ 8,694$$
$$25 \times \$\ 57.96 = \underline{+\ 1,449}$$
$$\$10,143$$

(f) Total handling cost = 125%(\$8,280) = \$10,350

Operating loss = Total handling cost (\$10,350) – Total selling price (\$10,143) = \$207

(g)
$$\_\_\% \ of \ C = OL$$
$$\_\_\%(\$8,280) = \$207$$
$$8,280x = 207$$
$$x = 2.5\%$$

Section 1

1. (a) P = $24,000   I = Prt   M = P + I
      r = 10%   = $24,000 × 0.10 × 1   = $24,000 + $2,400
      t = 1 year   I = $2,400   M = $26,400

   (b) P = $600   I = Prt   M = P + I
      r = 10%   = $600 × 0.10 × 0.5   = $600 + $30
      $t = \dfrac{6}{12} = 0.5$ I = $30   M = $630

   (c) P = $7,000   I = Prt   M = P + I
      r = 8%   = $57,000 + $1,520
      $t = \dfrac{4}{12} = \dfrac{1}{3}$ $= \$7,000 \times 0.08 \times \dfrac{1}{3}$ M = $58,520

                          I = $1,520

   (d) P = $800   I = Prt   M = P + I
      r = 8.5%   = $800 × 0.085 × 0.25   = $800 + $17
      $t = \dfrac{3}{12} = 0.25$ I = $17   M = $817

   (e) P = $2,400   I = Prt   M = P + I
      r = 9%   = $2,400 × 0.09 × 0.75   = $2,400 + $162
      $t = \dfrac{9}{12} = 0.75$ I = $162   M = $2,562

3. (a) $M = P(1 + rt)$ I = M - P
                                     = $26,400 - $24,000
      $= \$24,000(1 + 0.10 \cdot 1)$ I = $2,400

      $= 24,000(1 + 0.10)$

      $= 24,000(1.10)$

      $M = \$26,400$

   (c) $M = P(1 + rt)$ I = M - P
                                     = $58,520 - $57,000
      $= \$57,000\left(1 + \dfrac{8}{100} \cdot \dfrac{1}{3}\right)$ I = $1,520

      $= 57,000\left(1 + \dfrac{8}{300}\right)$

      $= 57,000\left(\dfrac{308}{300}\right)$

      $M = \$58,520$

3. (Continued)

(e) $M = P(1 + rt)$ $I = M - P$

$= \$2,400(1 + 0.09 \cdot 0.75)$ $= \$2,562 - \$2,400$

$= 2,400(1 + 0.0675)$ $I = \$162$

$= 2,400(1.0675)$

$M = \$2,562$

5.   P = \$690,000 $I = Prt$

    r = ? $\$36,225 = \$690,000 \times r \times 0.75$

    $t = \dfrac{9}{12} = 0.75$ $36,225 = 517,500r$

    I = \$36,225 $7\% = r$

7.   P = \$1,500 $I = Prt$

    r = 7% $\$87.50 = \$1,500 \times 0.07 \times t$

    t = ? $87.50 = 105t$

    I = \$1,587.50 - \$1,500 = \$87.50 $0.833\overline{3} = t$

                                $t = 0.8333 \times 12$ mos.

                                   = 10 mos.

9.   P = \$6,000 $I = Prt$

    r = ? $\$300 = \$6,000 \times r \times 0.5$

    $t = \dfrac{6}{12} = 0.5$ $300 = 3,000r$

    I = \$6,300 - \$6,000 = \$300 $10\% = r$

11.   P = ? $M = P(1 + rt)$

    r = 14% $\$104,060 = P(1 + 0.14 \cdot 1.5)$

    $t = \dfrac{18}{12} = 1.5$ $104,060 = P(1 + 0.21)$

                      $104,060 = P(1.21)$

    M = \$104,060 $\$86,000 = P$

13. Play the "What If" game. What if P = $1,000. Since the investment is to double at 16%, the interest must be $1,000.

P = $1,000   I = Prt

r = 16% $1,000 = $1,000 × 0.16 × t

t = ?   1,000 = 160t

I = $1,000   6.25 = t or $6\frac{1}{4}$ years

Section 2

1. (a) June   2 = 153 day   (b) Sept. 10 = 253 day
       March   6 = $\underline{-\ 65\ day}$ June   5 = $\underline{-156\ day}$
                    88 days   97 days

   (c) Dec. 10 = 344 day   (d) Nov. 20 = 324 day
       July 10 = $\underline{-191\ day}$ Jan. 14 = $\underline{-\ 14\ day}$
                   153 days   310 days

   (e) May 31, 2004 = 152 day   (f) Dec. 31, 2005 = 365 day
       Feb. 12, 2004 = $\underline{-\ 43\ day}$ Sept. 23, 2005 = $\underline{-266\ day}$
       (leap year)   109 days   99 days
                                         March 10, 2006 = $\underline{+\ 69\ day}$
                                                            168 days

   (g) Dec. 31, 2003 = 365 day   (h) Dec. 31, 2003 = 365 day
       July 4, 2003 = $\underline{-185\ day}$ Nov. 15, 2003 = $\underline{-319\ day}$
                        180 days   46 days
       March 4, 2004 = $\underline{+\ 63\ day}$ June 6, 2004 = $\underline{+158\ day}$
       (leap year)   243 days   (leap year)   204 days

3. (a) 232 day = Aug. 20   (b) 277 day = Oct. 4
        $\underline{+\ 30\ days}$ $\underline{+\ 60\ days}$
         262 day = Sept. 19   337 day = Dec. 3

   (c) 161 day = June 10   (d)   64 day = Mar. 5
        $\underline{+120\ days}$ $\underline{+200\ days}$
         281 day = Oct. 8   264 day = Sept. 21

   (e) 198 day = July 17   (f) 327 day = Nov. 23
        $\underline{+150\ days}$ $\underline{+180\ days}$
         348 day = Dec. 14   507 day
                                         $\underline{-365\ days}$
                                          142 day = May 22

5. (a) Due date = February 18 + 6 months = August 18

   August   18 =   230 day
   February 18 = - 49 day
                   181 days

   (b) Due date = May 24 + 2 months = July 24

   July   24 =   205 day
   May   24 = -144 day
                 61 days

   (c) Due date = January 26, 2006 + 9 months = October 26, 2006

   October 26 =   299 day
   January 26 = - 26 day
                  273 days

   (d) Due date = March 30, 2006 + 3 months = June 30, 2006

   June 30 =   181 day
   March 30 = - 89 day
                 92 days

   (e) Due date = January 13, 2004 + 5 months = June 13, 2004

   June   13 =   165 day
   January 13 = - 13 day
   (leap year)   152 days

Section 3

1. (a) <u>Ordinary interest</u>

   P = $45,000   I = Prt

   r = 8%   = $45,000 × 0.08 × 0.125

   $t = \frac{45}{360} = 0.125$ I = $450

   <u>Exact interest</u>

   P = $45,000   I = Prt

   r = 8% $= \$45,000 \times 0.08 \times \frac{45}{365}$

   $t = \frac{45}{365}$ I = $443.84

Business Mathematics: A Collegiate Approach

1. (Continued)

    (b) Ordinary interest

P = \$16,500

r = 9\%

$t = \dfrac{60}{360} = \dfrac{1}{6}$

I = Prt

$= \$16,500 \times 0.09 \times \dfrac{1}{6}$

I = \$247.50

Exact interest

P = \$16,500

r = 9\%

$t = \dfrac{60}{365}$

I = Prt

$= \$16,500 \times 0.09 \times \dfrac{60}{365}$

I = \$244.11

3. Due date = June 6 + 4 months = October 6

P = \$8,000

r = 14\%

Ordinary: $t = \dfrac{\text{Exact time}}{\text{Ordinary interest}} = \dfrac{122}{360}$

I = Prt

$= \$8,000 \times 0.14 \times \dfrac{122}{360}$

I = \$379.56

Exact: $t = \dfrac{\text{Exact time}}{\text{Exact interest}} = \dfrac{122}{365}$

I = Prt

$= \$8,000 \times 0.14 \times \dfrac{122}{365}$

I = \$374.36

5. P = \$3,000, $t = \dfrac{90}{360} = 0.25$ year

  (a) r = 5\%

I = Prt
= \$3,000 \times 0.05 \times 0.25
I = \$37.50

  (b) r = 10\%

I = Prt
= \$3,000 \times 0.10 \times 0.25
I = \$75

5. (Continued)

(c) r = 15%

I = Prt
= $3,000 × 0.15 × 0.25
I = $112.50

(d) r = 7.5%

I = Prt
= $3,000 × 0.075 × 0.25
I = $56.25

7. P = $4,000, r = 12%

(a) $t = \dfrac{90}{360} = 0.25$

I = Prt
= $4,000 × 0.12 × 0.25
I = $120

(b) $t = \dfrac{30}{360} = \dfrac{1}{12}$

I = Prt
= $4,000 × 0.12 × $\dfrac{1}{12}$
I = $40

(c) $t = \dfrac{120}{360} = \dfrac{1}{3}$

I = Prt
= $4,000 × 0.12 × $\dfrac{1}{3}$
I = $160

(d) $t = \dfrac{270}{360} = 0.75$

I = Prt
= $4,000 × 0.12 × 0.75
I = $360

9. At r = 8%, I = $36

(a) $\dfrac{16\%}{8\%} = 2;$ I = $36 × 2 = $72

(b) $\dfrac{10\%}{8\%} = \dfrac{5}{4};$ I = $36 × $\dfrac{5}{4}$ = $45

(c) $\dfrac{12\%}{8\%} = \dfrac{3}{2};$ I = $36 × $\dfrac{3}{2}$ = $54

11. At t = 120 days, I = $40

(a) $t = \dfrac{60}{120} = 0.5;$ I = $40 × 0.5 = $ 20

(b) $t = \dfrac{300}{60} = 5;$ I = $20 × 5   = $100

$\left(\text{or } t = \dfrac{300}{120} = 2.5;\ I = 40 × 2.5 = \$100\right)$

11. (Continued)

(c) $t = \dfrac{240}{120} = 2;$ $I = \$40 \times 2$ $= \$ 80$

13. (a) P = \$2,190   June 5 =   156 day
         r = 10%   April 5 = - 95 day
         $t = \dfrac{61}{365}$ 61 days

     I = Prt

       $= \$2,190 \times 0.10 \times \dfrac{61}{365}$

     I = \$36.60

(b) At 5%, $\dfrac{5}{10} = 0.5;$ $I = \$36.60 \times 0.5 = \$18.30$

(c) At 15%, $\dfrac{15}{10} = 1.5;$ $I = \$36.60 \times 1.5 = \$54.90$

15. (a) P = \$1,460   I = Prt

         r = 9.5% $= \$1,460 \times 0.095 \times \dfrac{150}{365}$

         $t = \dfrac{150}{365}$ I = \$57

(b) At t = 180 days, $\dfrac{180}{150} = \dfrac{5}{6};$ $I = \$57 \times \dfrac{5}{6} = \$ 68.40$

(c) At t = 270 days, $\dfrac{270}{150} = \dfrac{9}{5};$ $I = \$57 \times \dfrac{9}{5} = \$102.60$

Section 4

1.   (a) \$5,000.00   (f) \$5,000.00

           (b) William J. McDaniel   (g) 9%

           (c) Cardinal Bank   (h) 30 days

           (d) May 1, 20xx   (i) \$37.50

           (e) May 31, 20xx   (j) \$5,037.50

3.

	Principal	Rate	Date	Due Date	Time	Interest	Maturity Value
(a)	$4,800	9%	2/12	11/9	270 days	$324.00	$5,124.00
(b)	26,000	7	4/22	6/21	60 days	303.33	26,303.33
(c)	4,500	6	8/21	11/21	3 mos.	69.00	4,569.00
(d)	800	10	4/4	7/13	100 days	20.00	820.00

(a) November 9 = 313 day

February 12 = - 43 day

270 days

$I = Prt$ $M = P + I$

$= \$4,800 \times 0.09 \times 0.75$ $= \$4,800 + \$324$

$I = \$324$ $M = \$5,124$

(b) April 22 = 112 day

+ 60 days

Due date = 172 day = June 21

$I = Prt$ $M = P + I$

$= \$26,000 + \$303.33$

$= \$26,000 \times 0.07 \times \dfrac{1}{6}$ $M = \$26,303.33$

$I = \$303.33$

(c) Due date = August 21 + 3 months = November 21

November 21 = 325 day

August   21 = -233 day

92 days

3. (c) (Continued)

I = Prt

$$= \$4{,}500 \times 0.06 \times \frac{92}{360}$$

I = \$69

M = P + I

$$= \$4{,}500 + \$69$$

M = \$4,569

(d) July 13   =   194 day

        $\underline{-100\ days}$

Due date =   94 day   = April 4

I = Prt

$$\$20 = P \times 0.09 \times \frac{5}{18}$$

$$20 = 0.025P$$

$$\$800 = P$$

M = P + I

$$= \$800 + \$20$$

M = \$820

5. May   24, 2003 =   144 day

January 24, 2003 = $\underline{-\ 24\ day}$

                  120 days

I = Prt

$$= \$6{,}000 \times 0.09 \times \frac{120}{365}$$

I = \$177.53

P = \$6,000

r = 9%

$$t = \frac{120}{365}$$

M = P + I

$$= \$6{,}000.00 + \$177.53$$

M = \$6,177.53

7. Due date = December 5, 2000 = 340 day (leap year)

              October 5, 2000 = <u>-279 day</u>

                                    61 days

$t = \dfrac{61}{360}$

$I = Prt$ $M = P + I$

                                       $= \$16,500.00 + \$293.56$

$= \$16,500 \times 0.105 \times \dfrac{61}{360}$ $M = \$16,793.56$

$I = \$293.56$

9. $t = \dfrac{10}{12} = \dfrac{5}{6}$ $I = Prt$ $M = P + I$

                                           $= \$2,000 + \$150$

                $= \$2,000 \times 0.09 \times \dfrac{5}{6}$ $M = \$2,150$

                 $I = \$150$

11. August   14 = 226 day $I = \$2,298.60 - \$2,190.00$

    February 14 = <u>- 45 day</u> $= \$108.60$

                   181 days

    $t = \dfrac{181}{365}$

               $I = Prt$

  $\$108.60 = \$2,190 \times r \times \dfrac{181}{365}$

    $108.60 = 1,086r$

      $10\% = r$

13. $I = Prt$ $t = \dfrac{18}{73} \times 365$ days $= 90$ days

   $\$18 = \$730 \times 0.10 \times t$

    $18 = 73t$

    $\dfrac{18}{73} = t$

15.   I = Prt $t = \dfrac{36}{48} = 0.75;\ 0.75 \times 12$ mos. $= 9$ mos.

    $\$36 = \$400 \times 0.12 \times t$ or 0.75×360 days $= 270$ days

     $36 = 48t$

    $\dfrac{36}{48} = t$

17. P = ?   I = Prt

                        $\$105 = P \times 0.075 \times 0.5$

    r = 7.5%   105 $= 0.0375P$

                    $\$2{,}800 = P$

    $t = \dfrac{180}{360} = 0.5$

    I = \$105

19. December 6 = 340 day

    September 7 = <u>-250 day</u>

                  90 days

    P = ? $M = P(1 + rt)$

                 $\$11{,}275 = P(1 + 0.10 \cdot 0.25)$

                 11,275 $= P(1.025)$

    r = 10% $\$11{,}000 = P$

    $t = \dfrac{90}{360} = 0.25$

    M = \$11,275

Section 5

1.

	Maturity Value	Rate	Time	Present Value
(a)	$ 735	12%	150 days	$ <u>700</u>
(b)	2,484	14	90 days	<u>2,400</u>

1. (Continued)

(a) $M = P(1 + rt)$

$735 = P\left(1 + 0.12 \cdot \dfrac{5}{12}\right)$

$735 = P(1 + 0.05)$

$735 = P(1.05)$

$700 = P$

(b) $M = P(1 + rt)$

$2,484 = P(1 + 0.14 \cdot 0.25)$

$2,484 = P(1 + 0.035)$

$2,484 = P(1.035)$

$2,400 = P$

3.

	Principal	Rate	Time	Maturity Value	Rate Money is Worth	Present Value
(a)	$6,000	10%	90 days	$6,150	8%	$6,029.41
(b)	8,000	12	120 days	8,320	14	7,949.05

(a) $I = Prt$

$= \$6,000 \times 0.10 \times 0.25$

$I = \$150$

$M = P + I$

$= \$6,000 + \$150$

$M = \$6,150$

$M = P(1 + rt)$

$\$6,150 = P(1 + 0.08 \cdot 0.25)$

$6,150 = P(1 + 0.02)$

$6,150 = P(1.02)$

$\$6,029.41 = P$

(b) $I = Prt$

$= \$8,000 \times 0.12 \times \dfrac{1}{3}$

$I = \$320$

$M = P + I$

$= \$8,000 + \$320$

$M = \$8,320$

3. (b) (Continued)

$$M = P(1 + rt)$$

$$\$8,320 = P\left(1 + 0.14 \cdot \frac{1}{3}\right)$$

$$8,320 = P(1 + 0.04666666)$$

$$8,320 = P(1.04666666)$$

$$\$7,949.05 = P$$

5.

	Principal	Rate	Time	Maturity Value	Rate Money is Worth	Days Before Maturity Date	Present Value
(a)	$9,000	12%	300 days	$9,900	10%	180 days	$9,428.57
(b)	7,500	15	120 days	7,875	12	60 days	7,720.59

(a) I = Prt   M = P + I

                                        = $9,000 + $900

$$= \$9,000 \times 0.12 \times \frac{5}{6} \qquad M = \$9,900$$

I = $900

$$M = P(1 + rt)$$

$$\$9,900 = P(1 + 0.10 \cdot 0.5)$$

$$9,900 = P(1 + 0.05)$$

$$9,900 = P(1.05)$$

$$\$9,428.57 = P$$

(b) I = Prt   M = P + I

                                        = $7,500 + $375

$$= \$7,500 \times 0.15 \times \frac{1}{3} \qquad M = \$7,875$$

I = $375

5. (b) (Continued)

$$M = P(1 + rt)$$

$$\$7,875 = P\left(1 + 0.12 \cdot \frac{1}{6}\right)$$

$$7,875 = P(1 + 0.02)$$

$$7,875 = P(1.02)$$

$$\$7,720.59 = P$$

7.   September 15 = 258 day

    July   15 = <u>-196 day</u>

                  62 days

P = ?

$$M = P(1 + rt)$$

$$\$4,062 = P\left(1 + 0.09 \cdot \frac{62}{360}\right)$$

r = 9%

$$4,062 = P(1 + 0.0155)$$

$$4,062 = P(1.0155)$$

$$t = \frac{62}{360}$$

$$\$4,000 = P$$

M = \$4,062

9. April   15 = 105 day

   January 15 = <u>- 15 day</u>

              90 days

P = ?

$$M = P(1 + rt)$$

r = 11%

$$\$6,165 = P(1 + 0.11 \cdot 0.25)$$

$$t = \frac{90}{360} = 0.25$$

$$6,165 = P(1 + 0.0275)$$

$$6,165 = P(1.0275)$$

$$\$6,000 = P$$

M = \$6,165

11. P = \$3,500   I = Prt   M = P + I

                                                                      = \$3,500 + \$43.75

     r = 15%   = \$3,500 × 0.15 × $\dfrac{1}{12}$ M = \$3,543.75

     t = $\dfrac{30}{360}$ = $\dfrac{1}{12}$ I = \$43.75

     M = ?

     P = ? (present value) $M = P(1 + rt)$

                                        $\$3{,}543.75 = P\left(1 + 0.12 \cdot \dfrac{1}{12}\right)$

     r = 12% $3{,}543.75 = P(1 + 0.01)$

                                         $3{,}543.75 = P(1.01)$

     t = $\dfrac{1}{12}$ $\$3{,}508.66 = PV$

     M = \$3,543.75

13. June   12 = 163 day
     February 12 = $\underline{-\ 43\ day\ }$
                    120 days
     P = \$6,000   I = Prt   M = P + I

                                                                      = \$6,000 + \$220

     r = 11%   = \$6,000 × 0.11 × $\dfrac{1}{3}$ M = \$6,220

     t = $\dfrac{120}{360}$ = $\dfrac{1}{3}$ I = \$220

13. (Continued)

M = ?

P = ? (present value)

r = 12%

$$t = \frac{1}{3}$$

M = $6,220

$$M = P(1 + rt)$$

$$\$6,220 = P\left(1 + 0.12 \cdot \frac{1}{3}\right)$$

$$6,220 = P(1 + 0.04)$$

$$6,220 = P(1.04)$$

$$\$5,980.77 = PV$$

15. August 16 = 228 day

  May   16 = −136 day

              92 days

P = $12,000   I = Prt   M = P + I

r = 15% $= \$12,000 \times 0.15 \times \dfrac{92}{360}$ = $12,000 + $460

$t = \dfrac{92}{360}$ I = $460   M = $12,460

M = ?

August 16 = 228 day

July   17 = −198 day

              30 days

P = ? (present value)

r = 12%

$$t = \frac{30}{360} = \frac{1}{12}$$

M = $12,460

$$M = P(1 + rt)$$

$$\$12,460 = P\left(1 + 0.12 \cdot \frac{1}{12}\right)$$

$$12,460 = P(1 + 0.01)$$

$$12,460 = P(1.01)$$

$$\$12,336.63 = PV$$

17. May   17 =   137 day
     January 17 = – 17 day
                    120 days

P = $2,500   I = Prt   M = P + I

                                                   = $2,500 + $75

r = 9%   = $2,500 × 0.09 × $\frac{1}{3}$ M = $2,575

$t = \frac{120}{360} = \frac{1}{3}$ I = $75

M = ?

May   17 =   137 day

April 17 = –107 day

              30 days

P = ? (present value) $M = P(1 + rt)$

                           $2,575 = P\left(1 + 0.075 \cdot \frac{1}{12}\right)$

r = 7.5% $2,575 = P(1 + 0.00625)$

                           $2,575 = P(1.00625)$

$t = \frac{30}{360} = \frac{1}{12}$ $2,559.01 = PV$

M = $2,575

19. $30,000 cash now: $30,700 in $31,500 in
                             3 months: 6 months:

PV = $30,000 $PV = \frac{M}{(1 + rt)}$ $PV = \frac{M}{(1 + rt)}$

                     $= \frac{\$30,700}{(1 + 0.12 \cdot 0.25)}$ $= \frac{\$31,500}{(1 + 0.12 \cdot 0.5)}$

                     $= \frac{30,700}{(1 + 0.03)}$ $= \frac{31,500}{(1 + 0.06)}$

                     $= \frac{30,700}{(1.03)}$ $= \frac{31,500}{(1.06)}$

                     $PV = \$29,805.83$ $PV = \$29,716.98$

19. (Continued)

$30,000 cash now is the best offer because it has the highest present value.

CHAPTER 16 - BANK DISCOUNT
PROBLEM SOLUTIONS

Section 1

1. (a) $2,400
    (c) North Star Savings Assoc.
    (e) March 31, 20xx
    (g) 60 days
    (i) $2,360

    (b) A. Sample Borrower
    (d) January 30, 20xx
    (f) 10%
    (h) $40
    (j) $2,400

3.

	Maturity Value	Discount Rate	Date	Due Date	Time	Bank Discount	Proceeds
(a)	$9,000	8%	10/5	<u>11/4</u>	30 days	$ <u>60</u>	$<u>8,940</u>
(b)	38,500	7	2/25	<u>11/22</u>	270 days	<u>2,021.25</u>	<u>36,478.75</u>
(c)	5,000	10	5/14	8/12	<u>90 days</u>	<u>125</u>	<u>4,875</u>
(d)	3,000	9	<u>7/5</u>	9/3	60 days	<u>45</u>	<u>2,955</u>

(a)
$$\text{October 5} = 278 \text{ day}$$
$$\underline{+\ 30 \text{ days}}$$
$$\text{Due date} = 308 \text{ day} = \text{November 4}$$

$M = \$9,000$ $D = Mdt$ $p = M - D$
$$= \$9,000 - \$60$$
$d = 8\%$ $= \$9,000 \times 0.08 \times \dfrac{1}{12}$ $p = \$8,940$

$t = \dfrac{30}{360} = \dfrac{1}{12}$ $D = \$60$

$D = ?$

$p = ?$

(b)
$$\text{February 25} = 56 \text{ day}$$
$$\underline{+270 \text{ days}}$$
$$\text{Due date} = 326 \text{ day} = \text{November 22}$$

$M = \$38,500$ $D = Mdt$ $p = M - D$
$$= \$38,500 - \$2,021.25$$
$d = 7\%$ $= \$38,500 \times 0.07 \times \dfrac{3}{4}$ $p = \$36,478.75$

$t = \dfrac{270}{360} = \dfrac{3}{4}$ $D = \$2,021.25$

$D = ?$

$p = ?$

. (Continued)

(c) August 12 = 224 day
　　 May　　14 = $\underline{-134\ day}$
　　　　　　　　 90 days

　　 M = $5,000　　　D = Mdt　　　　　　　　p = M − D
　　　　　　　　　　　　　　　　　　　　　　　　= $5,000 − $125
　　 d = 10%　　　　　= $5,000 × 0.10 × $\frac{1}{4}$　　p = $4,875
　　 t = $\frac{90}{360}$ = $\frac{1}{4}$　　　D = $125
　　 D = ?

　　 p = ?

(d) September 3 = 246 day
　　　　　　　　　 $\underline{-\ 60\ days}$
　　 Date　　　　= 186 day = July 5

　　 M = $3,000　　　D = Mdt　　　　　　　　p = M − D
　　　　　　　　　　　　　　　　　　　　　　　　= $3,000 − $45
　　 d = 9%　　　　　= $3,000 × 0.09 × $\frac{1}{6}$　　p = $2,955
　　 t = $\frac{60}{360}$ = $\frac{1}{6}$　　　D = $45
　　 D = ?

　　 p = ?

. (a) M = $3,650　　　D = Mdt　　　　　　　　p = M − D
　　　　　　　　　　　　　　　　　　　　　　　　= $3,650 − $27
　　 d = 9%　　　　　= $3,650 × 0.09 × $\frac{30}{365}$　　p = $3,623
　　 t = $\frac{30}{365}$　　　　D = $27
　　 D = ?

　　 p = ?

5. (Continued)

(b) M = $7,300   D = Mdt   p = M - D
                                                          = $7,300 - $264
    d = 11%   = $7,300 × 0.11 × $\frac{120}{365}$ p = $7,036

    t = $\frac{120}{365}$ D = $264

    D = ?

    p = ?

(c) M = $1,095   D = Mdt   p = M - D
                                                          = $1,095 - $27
    d = 10%   = $1,095 × 0.10 × $\frac{90}{365}$ p = $1,068

    t = $\frac{90}{365}$ D = $27

    D = ?

    p = ?

7.

	Mat. Value	Disc. Rate	Time	Discount	Proceeds	Eq. Sim. Int. Rt.
(a)	$4,000	12%	90 days	$120	$3,880	12.5 %
(b)	65,000	8	150	2,166.67	62,833.33	8.25
(c)	126,000	7	120	2,940	123,060	7.25
(d)	5,400	10	60	90	5,310	10.25

(a) D = Mdt   p = M - D
                                        = $4,000 - $120
       = $4,000 × 0.12 × $\frac{1}{4}$ p = $3,880

    D = $120

            I = Prt

    $120 = $3,880 · r · $\frac{1}{4}$

      120 = 970r

    0.1237113 = r

      12.5% = r (nearest 1/4%)

. (Continued)

(b) D = Mdt

   $= \$65,000 \times 0.08 \times \dfrac{5}{12}$

   D = \$2,166.67

      I = Prt

$\$2,166.67 = \$62,833.33 \cdot r \cdot \dfrac{5}{12}$

 2,166.67 = 26,180.55r

0.0827587 = r

   8.25% = r (nearest $\dfrac{1}{4}$%)

p = M - D
  = \$65,000 - \$2,166.67
p = \$62,833.33

(c) D = Mdt

   $= \$126,000 \times 0.07 \times \dfrac{1}{3}$

   D = \$2,940

      I = Prt

   $\$2,940 = \$123,060 \cdot r \cdot \dfrac{1}{3}$

   2,940 = 41,020r

0.0716723 = r

   7.25% = r (nearest 1/4%)

p = M - D
  = \$126,000 - \$2,940
p = \$123,060

(d) D = Mdt

   $= \$5,400 \times 0.10 \times \dfrac{1}{6}$

   D = \$90

p = M - D
  = \$5,400 - \$90
p = \$5,310

7. (d) (Continued)

$$I = Prt$$

$$\$90 = \$5{,}310 \cdot r \cdot \frac{1}{6}$$

$$90 = 885r$$

$$0.1016949 = r$$

$$10.25\% = r \text{ (nearest 1/4\%)}$$

9.

	Maturity Value	Discount Rate	Date	Due Date	Time	Proceeds
(a)	$4,320	14%	1/9	6/9	5 mos.	$4,066.32
(b)	6,000	9	10/3	12/3	2 mos.	5,908.50
(c)	6,500	12	3/17	6/15	90 days	6,305.00
(d)	9,000	10	5/20	11/16	180 days	8,550.00

(a) $D = Mdt$

$$= \$4{,}320 \times 0.14 \times \frac{151}{360}$$

$$D = \$253.68$$

$p = M - D$
$\quad = \$4{,}320.00 - \$253.68$
$p = \$4{,}066.32$

(b) $D = Mdt$

$$= \$6{,}000 \times 0.09 \times \frac{61}{360}$$

$$D = \$91.50$$

$p = M - D$
$\quad = \$6{,}000.00 - \$91.50$
$p = \$5{,}908.50$

(c) $D = Mdt$
$\quad = \$6{,}500 \times 0.12 \times 0.25$
$D = \$195$

$p = M - D$
$\quad = \$6{,}500 - \$195$
$p = \$6{,}305$

(d) $D = Mdt$
$\quad = \$9{,}000 \times 0.10 \times 0.50$
$D = \$450$

$p = M - D$
$\quad = \$9{,}000 - \$450$
$p = \$8{,}550$

11. $M = \$46{,}000$

$d = 9\%$

$t = \dfrac{90}{360} = 0.25$

$D = Mdt$
$\quad = \$46{,}000 \times 0.09 \times 0.25$
$D = \$1{,}035$

$p = M - D$
$\quad = \$46{,}000 - \$1{,}035$
$p = \$44{,}965$

. D = $156   D = Mdt

  M = $7,800 $156 = $7,800 × 0.08 × t

  d = 8%   156 = 624t

  t = ? $\frac{1}{4}$ = t

                                 $\frac{1}{4}$ of 12 mos. = 3 mos. or

                                 90 days

. D = $9,400 - $8,554 = $846   D = Mdt

  M = $9,400 $846 = $9,400 × d × 0.75

  d = ?   846 = 7,050d

  t = $\frac{270}{360}$ = $\frac{3}{4}$ = 0.75   12% = d

. April   4 =   94 day
  January 4 = - 4 day
                 90 days

  M = $6,000   D = Mdt   p = M - D
                      = $6,000 × 0.13 × 0.25   = $6,000 - $195
  D = ?   D = $195   p = $5,805

  d = 13%

  t = $\frac{90}{360}$ = 0.25

  p = ?

.   M = $3,000
   d/r = 9%
     t = $\frac{6}{12}$ = 0.50

19. (Continued)

Simple interest note:
$$M = P(1 + rt)$$
$$\$3,000 = P(1 + .09 \cdot 0.50)$$
$$3,000 = P(1 + 0.045)$$
$$3,000 = P(1.045)$$
$$\$2,870.81 = P \text{ (present value)}$$

Simple discount note:
$$p = M(1 - dt)$$
$$= \$3,000(1 - .09 \cdot 0.5)$$
$$= 3,000(1 - 0.045)$$
$$= 3,000(0.9555)$$
$$p = \$2,865$$

21. $M = \$54,750$

$d = ?$

$t = \dfrac{60}{365}$

$D = \$54,750 - \$53,940 = \$810$

$p = \$53,940$

$$D = Mdt$$
$$\$810 = \$54,750 \times d \times \dfrac{60}{365}$$
$$810 = 9,000d$$
$$9\% = d$$

23. $M = \$10,000$

$d = 10\%$

$t = \dfrac{26 \times 7}{360} = \dfrac{182}{360} = \dfrac{91}{180}$

$D = ?$

$p = ?$

(b) $10,000

(a) $p = M(1 - dt)$

$$= \$10,000\left(1 - 0.10 \cdot \dfrac{91}{180}\right)$$
$$= 10,000(1 - 0.0505555)$$
$$= 10,000(0.9494445)$$
$$p = \$9,494.44$$

(c) $D = M - p$
$= \$10,000.00 - \$9,494.44$
$D = \$505.56$

Section 2

1.

	Proceeds	Discount Rate	Date	Due Date	Time	Maturity Value
(a)	$6,860	12%	10/18	12/17	60 days	$7,000
(b)	91,437.50	9	6/1	10/29	150 days	95,000
(c)	8,721	18	7/9	9/9	2 mos.	9,000
(d)	4,275	15	2/25	6/25	4 mos.	4,500

. (Continued)

(a) October 18 =   291 day
                  + 60 days
     Due date   =   351 day = December 17

$$p = M(1 - dt)$$

$$\$6,860 = M\left(1 - 0.12 \cdot \frac{1}{6}\right)$$

$$6,860 = M(1 - 0.02)$$

$$6,860 = M(0.98)$$

$$\$7,000 = M$$

(b) June 1   =   152 day
                 +150 days
     Due date =   302 day = October 29

$$p = M(1 - dt)$$

$$\$91,437.50 = M\left(1 - 0.09 \cdot \frac{5}{12}\right)$$

$$91,437.50 = M(1 - 0.0375)$$

$$91,437.50 = M(0.9625)$$

$$\$95,000 = M$$

(c) September 9 =   252 day
    July   9 = -190 day
                    62 days

$$p = M(1 - dt)$$

$$\$8,721 = M\left(1 - 0.18 \cdot \frac{31}{180}\right)$$

$$8,721 = M(1 - 0.031)$$

$$8,721 = M(0.969)$$

$$\$9,000 = M$$

1. (Continued)

(d) June   25 = 176 day
    February 25 = − 56 day
                   120 days

.

$$p = M(1 - dt)$$

$$\$4,275 = M\left(1 - 0.15 \cdot \frac{1}{3}\right)$$

$$4,275 = M(1 - 0.05)$$

$$4,275 = M(0.95)$$

$$\$4,500 = M$$

3.

	Original Note				Rediscounted Note			
	Maturity	Rate	Time	Proceeds	Rate	Time	Proceeds	Net Interest Earned
(a)	$7,600	13%	270 da	$6,859	14%	90 da	$7,334.00	$475.00
(b)	8,500	9	120 da	8,245	11	30 da	8,422.08	177.08
(c)	6,400	14	90 da	6,176	12	45 da	6,304.00	128.00

(a) Original Note: Rediscounted Note:
     D = $7,600 × 0.13 × 0.75   D = $7,600 × 0.14 × 0.25
     D = $741   D = $266

     p = $7,600 − $741   p = $7,600 − $266
     p = $6,859   p = $7,334

     Net interest earned = $7,334 − $6,859 = $475

(b) Original Note: Rediscounted Note:

     D = $8,500 × 0.09 × $\frac{1}{3}$ D = $8,500 × 0.11 × $\frac{1}{12}$

     D = $255   D = $77.92

     p = $8,500 − $255   p = $8,500.00 − $77.92
     p = $8,245   p = $8,422.08

     Net interest earned = $8,422.08 − $8,245.00 = $177.08

3. (Continued)

    (c)   <u>Original Note:</u>   <u>Rediscounted Note:</u>

$D = \$6{,}400 \times 0.14 \times 0.25$ $D = \$6{,}400 \times 0.12 \times \dfrac{1}{8}$

$D = \$224$ $D = \$96$

$p = \$6{,}400 - \$224$ $p = \$6{,}400 - \$96$
$p = \$6{,}176$ $p = \$6{,}304$

Net interest earned $= \$6{,}304 - \$6{,}176 = \$128$

5. (a)   <u>Simple Interest Note:</u>   <u>Discounted Note:</u>
            P = \$1,600   M = \$1,680

            r = 15%   d = 12%

$t = \dfrac{120}{360} = \dfrac{1}{3}$ $t = \dfrac{30}{360} = \dfrac{1}{12}$

$I = Prt$ $D = Mdt$

   $= \$1{,}600 \times 0.15 \times \dfrac{1}{3}$ $= \$1{,}680 \times 0.12 \times \dfrac{1}{12}$

$I = \$80$ $D = \$16.80$

$M = P + I$ $p = M - D$
  $= \$1{,}600 + \$80$ $= \$1{,}680.00 - \$16.80$
$M = \$1{,}680$ $p = \$1{,}663.20$

Net interest earned $= \$1{,}663.20 - \$1{,}600.00 = \$63.20$

    (b)   <u>Simple Interest Note:</u>   <u>Discounted Note:</u>
            P = \$26,500   M = \$27,825

            r = 12%   d = 10%

$t = \dfrac{150}{360} = \dfrac{5}{12}$ $t = \dfrac{45}{360} = \dfrac{1}{8}$

5. (b) (Continued)

$$I = Prt \qquad\qquad D = Mdt$$

$$= \$26,500 \times 0.12 \times \frac{5}{12} \qquad = \$27,825 \times 0.10 \times \frac{1}{8}$$

$$I = \$1,325 \qquad\qquad D = \$347.81$$

$$M = P + I \qquad\qquad p = M - D$$
$$= \$26,500 + \$1,325 \qquad = \$27,825 - \$347.81$$
$$M = \$27,825 \qquad\qquad p = \$27,477.19$$

Net interest earned = \$27,477.19 - \$26,500 = \$977.19

7. May   10 =   130 day
    January 10 = - 10 day
    _____
                120 days

$p = \$3,800$

$d = 15\%$

$t = \dfrac{120}{360} = \dfrac{1}{3}$

$M = ?$

$$p = M(1 - dt)$$

$$\$3,800 = M\left(1 - 0.15 \cdot \frac{1}{3}\right)$$

$$3,800 = M(1 - 0.05)$$

$$3,800 = M(0.95)$$

$$\$4,000 = M$$

9. July   20 =   201 day
    February 20 = - 51 day
    _____
                150 days

$p = \$8,470$

$d = 9\%$

$t = \dfrac{150}{360} = \dfrac{5}{12}$

$M = ?$

$$p = M(1 - dt)$$

$$\$8,470 = M\left(1 - 0.09 \cdot \frac{5}{12}\right)$$

$$8,470 = M(1 - 0.0375)$$

$$8,470 = M(0.9625)$$

$$\$8,800 = M$$

. June 3   =   154 day
                 + 90 days
Due date =   244 day = September 1

(a) M = $20,000   D = Mdt   p = M − D
     d = 16%   = $20,000 × 0.16 × 0.25   = $20,000 − $800
                         D = $800   p = $19,200
     t = $\frac{90}{360}$ = 0.25

     September 1   =   244 day
                       − 30 days
     Due date   =   214 day = August 2

(b) M = $20,000   D = Mdt   p = M − D

     d = 15%   = $20,000 × 0.15 × $\frac{1}{12}$ = $20,000 − $250

     t = $\frac{30}{360}$ = $\frac{1}{12}$ D = $250   p = $19,750

(c) Cullman National made $19,750 − $19,200 = $550

(d) $800 − $550 = $250 (or $20,000 − $19,750 = $250)

. (a) March 16 =   75 day
                    +120 days
        Due date =   195 day = July 14

       P = $9,000   I = Prt   M = P + I
                                                           = $9,000 + $420
       r = 14%   = $9,000 × 0.14 × $\frac{1}{3}$ M = $9,420

       t = $\frac{120}{360}$ = $\frac{1}{3}$ I = $420

13. (a) (Continued)

July 14 = 195 day
May 15 = -135 day
               60 days

M = $9,420   D = Mdt   p = M - D
                                                  = $9,420 - $188.40
d = 12%   = $9,420 × 0.12 × $\frac{1}{6}$ p = $9,231.60

t = $\frac{60}{360}$ = $\frac{1}{6}$ D = $188.40

(b) $9,231.60 - $9,000.00 = $231.60

(c) $420.00 - $231.60 = $188.40 (or $9,420.00 - $9,231.60 = $188.40)

15. (a) June 25 = 176 day
               +120 days
    Due date = 296 day = October 23

M = $6,300 D_1 = Mdt p_1 = M - D
                                                      = $6,300 - $210
d = 10%   = $6,300 × 0.10 × $\frac{1}{3}$ p_1 = $6,090

t = $\frac{120}{360}$ = $\frac{1}{3}$ D_1 = $210

(b) October 23 = 296 day
    July 25 = -206 day
               90 days

M = $6,300 D_2 = Mdt p_2 = M - D
                       = $6,300 × 0.09 × 0.25   = $6,300 - $141.75
d = 9% D_2 = $141.75 p_2 = $6,158.25

t = $\frac{90}{360}$ = 0.25

(c) Radford earned $6,158.25 - $6,090.00 = $68.25

(d) Radford lost $210.00 - $68.25 = $141.75

7. (a) May 13   =   133 day
                  + 90 days
        Due date = 223 day = Aug. 11

$P = \$52,000$

$r = 12\%$

$t = \dfrac{90}{360} = 0.25$

I = Prt
  = $52,000 × 0.12 × 0.25
I = $1,560

M = P + I
  = $52,000 + $1,560
M = $53,560

Aug. 11 = 223 day
June 22 = -173 day
            50 days

$M = \$53,560$

$d = 11\%$

$t = \dfrac{50}{365} = \dfrac{10}{73}$

D = Mdt

  $= \$53,560 \times 0.11 \times \dfrac{10}{73}$

D = $807.07

p = M - D
  = $53,560.00 - $807.07
p = $52,752.93

(b) $52,752.93 - $52,000.00 = $752.93

(c) $53,560 - $52,752.93 = $807.07

ection 3

	Amount of Invoice	Sales Terms	Amount Needed (Proceeds)	Time of Note	Discount Rate	Face Value of Note	Amount Saved
(a)	$16,500.00	3/10,N/30	$16,005.00	20 da	9%	$16,085.43	$414.57
(b)	7,614.80	2/15,N/30	7,462.50	15 da	12	7,500.00	114.80
(c)	4,242.86	2/15,N/30	4,158.00	45 da	8	4,200.00	42.86

1. (Continued)

(a)

$$\% \text{ Pd} \times L = N$$
$$0.97(\$16,500) =$$
$$\$16,005 = N$$

$$p = M(1 - dt)$$

$$\$16,005 = M\left(1 - 0.09 \cdot \frac{1}{18}\right)$$

$$16,005 = M(1 - 0.005)$$

$$16,005 = M(0.995)$$

$$\$16,085.43 = M$$

$$\begin{array}{ll} \$16,500.00 & \text{Amount of invoice} \\ \underline{-16,085.43} & \text{Face value of note} \\ \$\quad 414.57 & \text{Savings} \end{array}$$

(b)

$$\% \text{ Pd} \times L = N$$
$$0.98(\$7,614.80) =$$
$$\$7,462.50 = N$$

$$p = M(1 - dt)$$

$$\$7,462.50 = M\left(1 - 0.12 \cdot \frac{1}{24}\right)$$

$$7,462.50 = M(1 - 0.005)$$

$$7,462.50 = M(0.995)$$

$$\$7,500 = M$$

$$\begin{array}{ll} \$7,614.80 & \text{Amount of invoice} \\ \underline{-7,500.00} & \text{Face value of note} \\ \$\quad 114.80 & \text{Savings} \end{array}$$

(c)

$$\% \text{ Pd} \times L = N$$
$$0.98(\$4,242.86) =$$
$$\$4,158 = N$$

$$p = M(1 - dt)$$

$$\$4,158 = M\left(1 - 0.08 \cdot \frac{1}{8}\right)$$

$$4,158 = M(1 - 0.01)$$

$$4,158 = M(0.99)$$

$$\$4,200 = M$$

$$\begin{array}{ll} \$4,242.86 & \text{Amount of invoice} \\ \underline{-4,200.00} & \text{Face value of note} \\ \$\quad 42.86 & \text{Savings} \end{array}$$

. (a) $\quad$ % Pd × L = N $\qquad$ $p = M(1 - dt)$

$0.97(\$5,613.40) =$

$\$5,445 = N$ $\quad \$5,445 = M\left(1 - 0.18 \cdot \dfrac{1}{18}\right)$

$5,445 = M(1 - 0.01)$

$5,445 = M(0.99)$

$\$5,500 = M$

(b) $\qquad$ $\$5,613.40$ Amount of invoice

$\underline{-5,500.00}$ Face value of note

$\$\quad 113.40$ Savings

. (a) $\quad$ % Pd × L = N $\qquad$ $p = M(1 - dt)$

$0.97(\$3,073.45) =$

$\$2,981.25 = N$

$\$2,981.25 = M\left(1 - 0.15 \cdot \dfrac{1}{24}\right)$

$2,981.25 = M(1 - 0.00625)$

$2,981.25 = M(0.99375)$

$\$3,000 = M$

(b) $\$3,073.45$ Amount of invoice

$\underline{-3,000.00}$ Face value of note

$\$\quad 73.45$ Savings

. (a) $\qquad$ % Pd × L = N $\qquad$ $\$8,888.75$ Net cost of mdse.

$0.96(\$9,259.11) =$ $\qquad$ $\underline{+\quad 21.25}$ Freight

$\$8,888.75 = N$ $\qquad$ $\$8,910.00$ Proceeds

$p = M(1 - dt)$ $\qquad$ (b) $\$9,259.11$ Merchandise

$\underline{+\quad 21.25}$ Freight

$\$8,910 = M\left(1 - 0.09 \cdot \dfrac{1}{9}\right)$ $\qquad$ $\$9,280.36$ Amt. of invoice

$\underline{-9,000.00}$ Face value of note

$8,910 = M(1 - 0.01)$ $\qquad$ $\$\quad 280.36$ Savings

$8,910 = M(0.99)$

$\$9,000 = M$

9. (a)   % Pd × L = N
         0.98($2,715) =
             $2,660.70 = N

$2,660.70 Net cost of mdse.
+   18.30 Freight
$2,679.00 Proceeds

$$p = M(1 - dt)$$

$$\$2,679 = M\left(1 - 0.14 \cdot \frac{1}{18}\right)$$

$$2,679 = M(1 - 0.007778)$$

$$2,679 = M(0.992222)$$

$$\$2,700 = M$$

(b) $2,715.00 Merchandise
+   18.30 Freight
$2,733.30 Amt. of invoice
-2,700.00 Face value of note
$ 33.30 Savings

Section 4

1. P = $22,500

    r =   9%

    $$t = \frac{60}{360} = \frac{1}{6}$$

    I = Prt

    $$= \$22,500 \times 0.09 \times \frac{1}{6}$$

    I = $337.50

    M = P + I
      = $22,500 + $337.50
    M = $22,837.50

3. M = $16,000

    d = 10%

    $$t = \frac{60}{360} = \frac{1}{6}$$

    D = Mdt

    $$= \$16,000 \times 0.10 \times \frac{1}{6}$$

    D = $266.67

    p = M - D
      = $16,000 - $266.67
    p = $15,733.33

5. (a) Interest Note:
         P = $1,600

         r = 15%

         $$t = \frac{120}{360} = \frac{1}{3}$$

         Discount Note:
         M = $1,600

         d = 15%

         $$t = \frac{120}{360} = \frac{1}{3}$$

5. (a) (Continued)

Interest Note: Discount Note:
I = Prt   D = Mdt

$\quad = \$1,600 \times 0.15 \times \dfrac{1}{3}$ $\quad = \$1,600 \times 0.15 \times \dfrac{1}{3}$

I = \$80   D = \$80

(b) Amount received: Interest note = \$1,600
                       Discount note = \$1,600 - \$80 = \$1,520

(c) Maturity value: Interest note = \$1,600 + \$80 = \$1,680
                       Discount note = \$1,600

7. (a) p = \$2,352

$\quad$ d = 12%

$\quad t = \dfrac{60}{360} = \dfrac{1}{6}$

$\quad$ M = ?

$$M = \frac{p}{1 - dt}$$

$$= \frac{\$2,352}{1 - 0.12 \cdot \dfrac{1}{6}}$$

$$= \frac{2,352}{1 - 0.02}$$

$$= \frac{2,352}{0.98}$$

$$M = \$2,400$$

(b) \$2,400 = Maturity value = face value

9. (a) M = \$16,940

$\quad$ r = 8%

$\quad t = \dfrac{120}{360} = \dfrac{1}{3}$

$\quad$ P = ?

$$M = P(1 + rt)$$

$$\$16,940 = P(1 + 0.08 \cdot \tfrac{1}{3})$$

$$16,940 = P(1 + 0.0266666)$$

$$16,940 = P(1.0266666)$$

$$\$16,500 = P$$

(b) The maker received the face value (principal) = \$16,500.

1. $M = \$7,000$

   $d = 10\%$

   $t = \dfrac{90}{360} = 0.25$

   $r = ?$

   $D = Mdt$

   $\quad = \$7,000 \times 0.10 \times 0.25$

   $D = \$175$

   $p = M - D$
   $\quad = \$7,000 - \$175$
   $p = \$6,825$

   $\quad\quad I = Prt$

   $\quad \$175 = \$6,825 \cdot r \cdot 0.25$

   $\quad\quad 175 = 1,706.25r$

   $10.25\% = r$ (nearest 1/4%)

3. | Present value of a <u>Simple Interest Note</u>: | Proceeds of a <u>Bank Discount Note</u>: |
   |---|---|
   | $M = \$8,400$ | $M = \$8,400$ |
   | $r = 15\%$ | $d = 15\%$ |
   | $t = \dfrac{120}{360} = \dfrac{1}{3}$ | $t = \dfrac{120}{360} = \dfrac{1}{3}$ |
   | $P = ?$ | $p = ?$ |

   Present value of a
   <u>Simple Interest Note</u>:
   $$M = P(1 + rt)$$
   $$\$8,400 = P\left(1 + 0.15 \cdot \frac{1}{3}\right)$$
   $$8,400 = P(1 + .05)$$
   $$8,400 = P(1.05)$$
   $$\$8,000 = P$$

   Proceeds of a
   <u>Bank Discount Note</u>:
   $$p = M(1 - dt)$$
   $$= \$8,400\left(1 - 0.15 \cdot \frac{1}{3}\right)$$
   $$= 8,400(1 - 0.05)$$
   $$= 8,400(0.95)$$
   $$p = \$7,980$$

. December 22 = 356 day
August   22 = -234 day
                122 days

P = $1,080

r = 10%

$t = \dfrac{122}{360} = \dfrac{61}{180}$

I = Prt

$= \$1,080 \times 0.10 \times \dfrac{61}{180}$

I = $36.60

M = P + I
  = $1,080.00 + $36.60
M = $1,116.60

M = $1,116.60

r = 9%

$t = \dfrac{122}{360} = \dfrac{61}{180}$

$M = P(1 + rt)$

$\$1,116.60 = P\left(1 + 0.09 \cdot \dfrac{61}{180}\right)$

$1,116.60 = P(1 + 0.0305)$

$1,116.60 = P(1.0305)$

$\$1,083.55 = P$ (present value)

. July   5 = 186 day
February 5 = - 36 day
               150 days

P = $5,200

r = 12%

$t = \dfrac{150}{360} = \dfrac{5}{12}$

I = Prt

$= \$5,200 \times 0.12 \times \dfrac{5}{12}$

I = $260

M = P + I
  = $5,200 + $260
M = $5,460

17. (Continued)

M = $5,460

r = 10%

$t = \dfrac{90}{360} = 0.25$

$M = P(1 + rt)$

$5,460 = P(1 + 0.10 \cdot 0.25)$

$5,460 = P(1 + 0.025)$

$5,460 = P(1.025)$

$5,326.83 = P$ (present value)

19. (a) January 8 =   8 day
                      = +180 days
         July 7   =   188 day
         (Due date)

P = $9,500

r = 14%

$t = \dfrac{180}{360}$

I = Prt

$= \$9,500 \times 0.14 \times \dfrac{180}{360}$

I = $665

M = P + I
  = $9,500 + $665
M = $10,165

July 7 =   188 day
April 9 = − 99 day
                89 days

M = $10,165

d = 15%

$t = \dfrac{89}{360}$

$p = M(1 - dt)$

$= \$10,165 \left(1 - 0.15 \cdot \dfrac{89}{360}\right)$

$= 10,165(1 - 0.0370833)$

$= 10,165(0.9629167)$

$p = \$9,788.05$

(b) Cook Construction Co. made
     $9,788.05 − $9,500.00 =
     $288.05.

(c) The bank earned the discount
     $10,165.00 − $9,788.05 =
     $376.95.

(a) August   15 = 227 day
                    +120 day
    December 13 = 347 days
    (Due date)

M = $14,000

d = 9%

$t = \dfrac{120}{360} = \dfrac{1}{3}$

$p_1 = M(1 - dt)$

$= \$14,000\left(1 - 0.09 \cdot \dfrac{1}{3}\right)$

$= 14,000(1 - 0.03)$

$= 14,000(0.97)$

$p_1 = \$13,580$

(b) December 13 =   347 day
    October   16 = -289 day
                      58 days

M = $14,000

d = 8.5%

$t = \dfrac{58}{360} = \dfrac{29}{180}$

$p_2 = M(1 - dt)$

$= \$14,000\left(1 - 0.085 \cdot \dfrac{29}{180}\right)$

$= 14,000(1 - 0.0136944)$

$= 14,000(0.9863056)$

$p_2 = \$13,808.28$

(c) American Bank earned $13,808.28 - $13,580.00 = $228.28

(d) The second bank earned $14,000.00 - $13,808.28 = $191.72

Section 1

1. (a)

P = \$6,000	Original principal	\$6,000.00
r = 12%		
$t_1 = \dfrac{20}{360} = \dfrac{1}{18}$	1st partial payment	\$2,040
	Less: int. (20 days)	$-$ 40
$I_1 = Prt$	Payment to principal	$\underline{-2,000.00}$
$= \$6,000 \times 0.12 \times \dfrac{1}{18}$	Adjusted principal	\$4,000.00
$I_1 = \$40$	2nd partial payment	\$1,500
	Less: int. (60 days)	$-$ 80
$t_2 = \dfrac{60}{360} = \dfrac{1}{6}$	Payment to principal	$\underline{-1,420.00}$
	Adjusted principal	\$2,580
$I_2 = Prt$	Interest due (70 days)	$+$ 60.20
$= \$4,000 \times 0.12 \times \dfrac{1}{6}$	Balance due	\$2,640.20
$I_2 = \$80$		

$$t_3 = \frac{70}{360}$$

$I_3 = Prt$

$\quad = \$2,580 \times 0.12 \times \dfrac{7}{36}$

$I_3 = \$60.20$

1. (Continued)

    (b)

$P = \$4,500$	Original principal $\$4,500.00$

$P = \$4,500$

$r = 8\%$

$t_1 = \dfrac{20}{360} = \dfrac{1}{18}$

$I_1 = Prt$

    $= \$4,500 \times 0.08 \times \dfrac{1}{18}$

$I_1 = \$20$

$t_2 = \dfrac{45}{360} = \dfrac{1}{8}$

$I_2 = Prt$

    $= \$2,490 \times 0.08 \times \dfrac{1}{8}$

$I_2 = \$24.90$

$t_3 = \dfrac{25}{360} = \dfrac{5}{72}$

$I_3 = Prt$

    $= \$1,484.90 \times 0.08 \times \dfrac{5}{72}$

$I_3 = \$8.25$

Original principal		$\$4,500.00$
1st partial payment	$\$2,030$	
Less: int. (20 days)	$-\ \ \ 20$	
Payment to principal		$-2,010.00$
Adjusted principal		$\$2,490.00$
2nd partial payment	$\$1,030$	
Less: int. (45 days)	-24.90	
Payment to principal		$-1,005.10$
Adjusted principal		$\$ 1484.90$
Interest due (25 days)		8.25
Balance due		$\$1,493.15$

1. (Continued)

   (c)

$P = \$8,000$ Original principal $\$8,000$

$r = 10\%$

$t_1 = \dfrac{90}{360} = \dfrac{1}{4}$ 1st partial payment $\$2,200$

                 Less: int. (90 days) $-\underline{\quad 200}$

$I_1 = Prt$ Payment to principal $\underline{-2,000}$

     $= \$8,000 \times 0.10 \times \dfrac{1}{4}$ Adjusted principal $\$6,000$

$I_1 = \$200$ 2nd partial payment $\$2,875$

                 Less: int. (45 days) $-\underline{\quad 75}$

$t_2 = \dfrac{45}{360} = \dfrac{1}{8}$ Payment to principal $\underline{-2,800}$

                 Adjusted principal $\$3,200$

$I_2 = Prt$ Interest due (135 days) $+\underline{\quad 120}$

     $= \$6,000 \times 0.10 \times \dfrac{1}{8}$ Balance due $\$3,320$

$I_2 = \$75$

$t_3 = \dfrac{135}{360} = \dfrac{3}{8}$

$I_3 = Prt$

     $= \$3,200 \times 0.10 \times \dfrac{3}{8}$

$I_3 = \$120$

3.

$P = \$45,000$

$r = 10\%$

$t_1 = \dfrac{60}{360} = \dfrac{1}{6}$

$I_1 = Prt$

$\quad = \$45,000 \times 0.10 \times \dfrac{1}{6}$

$I_1 = \$750$

$t_2 = \dfrac{90}{360} = \dfrac{1}{4}$

$I_2 = Prt$

$\quad = \$25,750 \times 0.10 \times \dfrac{1}{4}$

$I_2 = \$643.75$

$t_3 = \dfrac{30}{360} \times \dfrac{1}{12}$

$I_3 = Prt$

$\quad = \$11,393.75 \times 0.10 \times \dfrac{1}{12}$

$I_3 = \$94.95$

Original principal, 4/5		$10,000
1st partial payment, 6/4	$20,000	
Less: int. (60 days)	− 750	
Payment to principal		−19,250
Adjusted balance		$25,750
2nd partial payment, 9/2	$15,000	
Less: int. (90 days)	−643.75	
Payment to principal		−14,356.25
Adjusted principal		$11,393.75
Interest due (30 days)		+ 94.95
Balance due, 10/2		$11,488.70

5.

$P = \$12,000$

$r = 7\%$

$t_1 = \dfrac{20}{360} = \dfrac{1}{18}$

$I_1 = Prt$

$= \$12,000 \times 0.07 \times \dfrac{1}{18}$

$I_1 = \$46.67$

$t_2 = \dfrac{45}{360} = \dfrac{1}{8}$

$I_2 = Prt$

$= \$9,926.67 \times 0.07 \times \dfrac{1}{8}$

$I_2 = \$86.86$

$t_3 = \dfrac{25}{360} = \dfrac{5}{72}$

$I_3 = Prt$

$= \$4,788.53 \times 0.07 \times \dfrac{5}{72}$

$I_3 = \$23.28$

Original principal, 6/10	$12,000.00
1st partial pmt., 6/30 $2,120	
Less: int. (20 days)	-46.67
Payment to principal	- 2,073.33
Adjusted principal	$ 9,926.67
2nd partial pmt., 8/14 $5,225	
Less: int. (45 days)	-86.86
Payment to principal	- 5,138.14
Adjusted principal	$ 4,788.53
Interest due (25 days)	+ 23.28
Balance due, 9/8	$ 4,811.81

Section 2

1. (a)

Pmt. No.	Balance Due	Interest $(1\frac{1}{4}\%)$	Monthly Payment	Payment Toward Bal. Due	Adjusted Balance Due
1	$260.00	$3.25	$ 90.00	$ 86.75	$173.25
2	173.25	2.17	90.00	87.83	85.42
3	85.42	1.07	86.49	85.42	0.00
	Totals	$6.49	$266.49	$260.00	

The interest is $6.49.

(b)

Pmt. No.	Balance Due	Interest (1%)	Monthly Payment	Payment Toward Bal. Due	Adjusted Balance Due
1	$500.00	$ 5.00	$200.00	$195.00	$305.00
2	305.00	3.05	200.00	196.95	108.05
3	108.05	1.08	109.13	108.05	0.00
	Totals	$ 9.13	$509.13	$500.00	

The interest is $9.13.

3.

Mo.	Date	Prev. Balance	Payment	Adj. Bal.	Avg. Bal.	2% Int.	Purchases	Current Balance
1	18	$300.00	$ 50	$250.00				$250.00
	20						$20	270.00
	30				$280.00[1]	$ 5.60		275.60
2	15	275.60	75	200.60				200.60
	19						25	225.60
	25						10	235.60
	30				238.10[2]	4.76		240.36
3	12	240.36	45	195.36				195.36
	28						30	225.36
	30		____		213.36[3]	4.27	___	229.63
		Totals	$170			$14.63	$85	

$$(1) \quad \frac{(\$300 \times 18) + (\$250 \times 12)}{30} = \$280.00$$

$$(2) \quad \frac{(\$275.60 \times 15) + (\$200.60 \times 15)}{30} = \$238.10$$

$$(3) \quad \frac{(\$240.36 \times 12) + (\$195.36 \times 18)}{30} = \$213.36$$

Check:

$300.00 Beginning balance

+ 85.00 Purchases

+ 14.63 Interest

$399.63

-170.00 Payments

$229.63 Ending balance

5.

Pmt. No.	Balance Due	Interest (1%)	Monthly Payment	Payment Toward Bal. Due	Adjusted Balance Due
1	$459.00	$ 4.59	$100.00	$ 95.41	$363.59
2	363.59	3.64	100.00	96.36	267.23
3	267.23	2.67	100.00	97.33	169.60
4	169.90	1.70	100.00	98.30	71.60
5	71.60	0.72	72.32	71.60	0.00
	Totals	$13.32	$472.32	$459.00	

The interest will total $13.32

7.

Mo.	Date	Prev. Balance	Payment	Adj. Bal.	Avg. Bal.	1% Int.	Purchases	Current Balance
1	15	$ 85.00	$10	$ 75.00				$ 75.00
	20						$ 30	105.00
	30				$ 80.00[1]	$0.80		105.80
2	18	105.80	20	85.80				85.80
	25						45	130.80
	30				97.80[2]	0.98		131.78
3	12	131.78	20	111.78				111.78
	21						25	136.78
	30		___		119.78[3]	1.20	___	137.98
		Totals	$50			$2.98	$100	

$$(1) \quad \frac{(\$85 \times 15) + (\$75 \times 15)}{30} = \$80.00$$

7. (Continued)

(2) $\dfrac{(\$105.80 \times 18) + (\$85.80 \times 12)}{30} = \$97.80$

(3) $\dfrac{(\$131.78 \times 12) + (\$111.78 \times 18)}{30} = \$119.78$

Check:

$ 85.00	Beginning balance	(a)	Total payments = $ 50.00
+100.00	Purchases	(b)	Total interest = $ 2.98
+ 2.98	Interest	(c)	Ending balance = $137.98
$187.98			
- 50.00	Payments		
$137.98	Ending balance		

Section 3

1. (a)

Cash price	$500
Down payment	- 50
Outstanding balance	$450
Finance charge	+ 20
Total of payments	$470

$\dfrac{\$470}{10} = \$47/\text{month}$

(b)

Cash price	$1,800
Down payment (25%)	- 450
Outstanding balance	$1,350
Finance charge	+ 57
Total of payments	$1,407

$\dfrac{\$1,407}{12} = \$117.25/\text{month}$

(c)

Cash price	$595
Down payment (10%)	- 59.50
Outstanding balance	$535.50
Finance charge	+ 26.78 (= $535.50 × 0.10 × 0.5)
Total of payments	$562.28

$\dfrac{\$562.28}{6} = \$93.71/\text{month}$

3. (a) $\dfrac{\text{Finance charge}}{\text{Amount financed}} \times 100 = \dfrac{\$20}{\$120} \times 100 = \$16.67$

$16.67 finance charge per $100 financed over 12 months = 29.5% actuarial rate

(b) $\dfrac{\text{Finance charge}}{\text{Amount financed}} \times 100 = \dfrac{\$30}{\$300} \times 100 = \$10$

$10 finance charge per $100 financed over 9 months = 23.5% actuarial rate

(c) Simple interest rate × time = 14% × 3 = 42

$42 interest per $100 financed over 36 months = 24.5% actuarial rate

(d) Simple interest rate × time = $12\% \times \dfrac{5}{12} = 5$

$5 interest per $100 financed over 5 months = 19.75% actuarial rate

5.

Month	Prt	=	I	Payment to Principal
1	$800.00 × 0.17 × $\frac{1}{12}$	=	$11.33	$128.67
2	671.33 × 0.17 × $\frac{1}{12}$	=	9.51	130.49
3	540.84 × 0.17 × $\frac{1}{12}$	=	7.66	132.34
4	408.50 × 0.17 × $\frac{1}{12}$	=	5.79	134.21
5	274.29 × 0.17 × $\frac{1}{12}$	=	3.89	136.11
6	138.18 × 0.17 × $\frac{1}{12}$	=	1.96	138.04
			$40.14	$799.86

($0.14 difference in total interest and total payment to principal is due to rounding interest and usage of an effective interest rate correct only to the nearest 1/4%.)

7. Cash price $289
   Down payment (1/4)   − 72.25
   Outstanding balance $216.75
   Finance charge   + 40
   Total of payments $256.75

$$\frac{\$256.75}{6} = \$42.79/month$$

9. Cash price $2,500
   Down payment (20%)   − 500
   Outstanding balance $2,000
   Finance charge   + 50
   Total of payments $2,050

$$\frac{\$2,050}{18} = \$113.89/month$$

11. (a) List price $520
   Down payment (25%)   −130
   Outstanding balance $390
   Finance charge   + 39 $\left(= \$390 \times 0.12 \times \frac{10}{12} \right)$
   Total owed $429

The finance charge = $39

(b) $$\frac{\$429}{10} = \$42.90/month$$

(c) Total payments $429 $520 List price
   Down payment   +130   or   + 39 Finance charge
   Total cost $559 $559 Total cost

(d) Simple interest rate × time = 12% × $\frac{10}{12}$ = 10 or

$$\frac{Finance\ charge}{Amount\ financed} \times 100 = \frac{\$39}{\$390} \times 100 = 10$$

$10 finance charge per $100 financed over 10 months
= 21.25% actuarial rate

13. (a) $125 × 24 payments = $3,000 Total payments

                              −2,600 Amount borrowed

                         $ 400 Finance charge

    (b) $$\frac{\text{Finance charge}}{\text{Amount financed}} \times 100 = \frac{\$400}{\$2,600} \times 100 = 15.38$$

            $15.38 per $100 financed over 24 months = 14.25% APR

15. (a) $35 × 18 payments = $630 Total payments

                              −580 List price

                         $ 50 Finance charge

    (b) $$\frac{\text{Finance charge}}{\text{Amount financed}} \times 100 = \frac{\$50}{\$580} \times 100 = 8.62$$

            $8.62 per $100 financed over 18 months = 10.5% APR

17. (a) $70 × 12 payments = $ 840 Total payments

                              +   200 Down payment

                         $1,040 Total cost when

                                    purchased on credit

    (b) $1,040 Total cost   (c) $895 Cash price

            −   895 Cash price   −200 Down payment

            $ 145 Finance charge $695 Amount financed

            $$\frac{\$145}{\$695} \times 100 = 20.86$$ $20.86 per $100 financed

                                          over 12 months = 36.5% APR

19. (a) $75 × 36 payments = $2,700 Total payments
                              + 400 Down payment
                           $3,100 Total cost on the
                                  installment plan

    (b) $3,100 Total cost   (c) $2,695 Cash price
         -2,695 Cash price   - 400 Down payment
         $ 405 Finance charge $2,295 Amount financed

         $$\frac{\$405}{\$2,295} \times 100 = 17.65$$ $17.65 per $100 financed
                                                  over 36 months = 11.0%
                                                  APR

21. <u>14% simple interest</u>

    (a) <u>12 monthly installments</u>
        Rate × time = 14 × 1 = 14
        $14 per $100 financed over 12 months = 25% annual rate

    (b) <u>18 monthly installments</u>
        Rate × time = $14 \times \dfrac{18}{12} = 21$
        $21 per $100 financed over 18 months = 25% annual rate

    (c) <u>9 monthly installments</u>

        Rate × time = $14 \times \dfrac{9}{12} = 10.5$
        $10.50 per $100 financed over 9 months = 24.5% annual rate

23. (a) $I = Prt$

$= \$2,400 \times 0.09 \times \dfrac{1}{2}$

$I = \$108$

(b) $M = P + I$

$\quad = \$2,400 + \$108$

$M = \$2,508$

$\dfrac{\$2,508}{6} = \$418/\text{month}$

(c) Simple interest × time = $9\% \times \dfrac{1}{2} = 4.5$

$4.50 per $100 financed over 6 months = 15.25% actuarial rate

(d)

Month	Prt	=	I	Payment to Principal
1	$\$2,400.00 \times 0.1525 \times \dfrac{1}{12}$	=	$ 30.50	$ 387.50
2	$2,012.50 \times 0.1525 \times \dfrac{1}{12}$	=	25.58	392.42
3	$1,620.08 \times 0.1525 \times \dfrac{1}{12}$	=	20.59	397.41
4	$1,222.67 \times 0.1525 \times \dfrac{1}{12}$	=	15.54	402.46
5	$820.21 \times 0.1525 \times \dfrac{1}{12}$	=	10.42	407.58
6	$412.63 \times 0.1525 \times \dfrac{1}{12}$	=	5.24	412.76
		Totals =	$107.87	$2,400.13

($0.13 difference in totals is due to rounding and use of the actuarial rate correct only to nearest 1/4%.)

25. (a) I = Prt$\qquad\qquad$(b) $\dfrac{\$2,310}{5}$ = \$462/month

$\qquad$ = \$2,200 × 0.12 × $\dfrac{5}{12}$

$\qquad$ I = \$110

(c) Simple interest rate × time = 12% × $\dfrac{5}{12}$ = 5

$\quad$ \$5 finance charge per \$100 financed over 5 months = 19.75% APR

(d)

Month	Prt	=	I	Payment to Principal
1	\$2,200.00 × 0.1975 × $\dfrac{1}{12}$	=	\$ 36.21	\$ 425.79
2	1,774.21 × 0.1975 × $\dfrac{1}{12}$	=	29.20	432.80
3	1,341.41 × 0.1975 × $\dfrac{1}{12}$	=	22.08	439.92
4	901.49 × 0.1975 × $\dfrac{1}{12}$	=	14.84	447.16
5	454.33 × 0.1975 × $\dfrac{1}{12}$	=	7.48	454.52
	Totals	=	\$109.81	\$2,200.19

(\$0.19 difference in totals is due to rounding and use of actuarial tables correct only to the nearest 1/4%.)

Section 4

1. (a) Value for 6 payments at 18.25% = 5.39

$$\frac{n \times Pmt \times Value}{100 + Value} = \frac{6 \times \$110 \times 5.39}{100 + 5.39}$$

$$= \frac{3,557.40}{105.39}$$

$$= \$33.75 \text{ Interest saved}$$

$660.00 Total of 6 remaining payments of $110 each

- 33.75 Interest saved

$626.25 Balance due

(b) Value for 5 payments at 26.75% = 6.79

$$\frac{n \times Pmt \times Value}{100 + Value} = \frac{5 \times \$237.50 \times 6.79}{100 + 6.79}$$

$$= \frac{8,063.125}{106.79}$$

$$= \$75.50 \text{ Interest saved}$$

$1,187.50 Total 5 remaining payments of $237.50 each

- 75.50 Interest saved

$1,112.00 Balance due

3. (a) <u>Rule of 78s</u>: 6 months early out of 24:

Numerator: Denominator:

$$\frac{6(6+1)}{2} = \frac{42}{2} = 21 \qquad\qquad \frac{24(24+1)}{2} = \frac{600}{2} = 300$$

$$\frac{21}{300} \times \$440 = \$30.80 \text{ Interest saved}$$

$660.00 Total of 6 remaining payments of $110 each
<u>- 30.80</u> Interest saved
$629.20 Balance due

(b) <u>Rule of 78s</u>: 5 months early out of 15:

Numerator: Denominator:

$$\frac{5(5+1)}{2} = \frac{30}{2} = 15 \qquad\qquad \frac{15(16)}{2} = \frac{240}{2} = 120$$

$$\frac{15}{120} \times \$562.50 = \$70.31 \text{ Interest saved}$$

$1,187.50 Total of 5 remaining payments of $237.50 each
<u>-   70.31</u> Interest saved
$1,117.19 Balance due

5. (a) $I = Prt$

$$= \$2,800 \times 0.10 \times \frac{24}{12}$$

$I = \$560$

(b) $M = P + I$

$$= \$2,800 + \$560$$

$M = \$3,360$

$$\frac{\$3,360}{24} = \$140/month$$

(c) $\dfrac{\$560}{\$2,800} \times 100 = 20$

$20.00 per $100 financed over 24 months = 18.25%

(d) $\dfrac{n \times Pmt \times Value}{100 + Value} = \dfrac{14 \times \$140 \times 11.78}{100 + 11.78}$

$$= \frac{23,088.80}{111.78}$$

$$= \$206.56 \text{ Interest saved}$$

(e) $1,960.00 Total of 14 remaining payments of $140 each

  − 206.56 Interest saved

$1,753.44 Balance due

7. (a) $I = Prt$

$$= \$7,500 \times 0.11 \times \frac{15}{12}$$

$I = \$1,031.25$

(b) $M = P + I$

$$= \$7,500.00 + \$1,031.25$$

$M = \$8,531.25$

$$\frac{\$8,531.25}{15} = \$568.75/month$$

(c) $\dfrac{\$1,031.25}{\$7,500} \times 100 = 13.75$

$13.75 per $100 financed over 15 months = 19.75%

7. (Continued)

(d) $\dfrac{n \times Pmt \times Value}{100 + Value} = \dfrac{6 \times \$568.75 \times 5.84}{100 + 5.84}$

$$= \dfrac{19,929}{105.84}$$

$$= \$188.29 \text{ Interest saved}$$

(e) $3,412.50 Total of remaining 6 payments of $568.75 each

  −   188.29 Interest saved

  $3,224.21 Balance due

9. (a) $I = Prt$

    $= \$3,600 \times 0.08 \times \dfrac{9}{12}$

    $I = \$216$

(b) $M = P + I$

    $= \$3,600 + \$216$

    $M = \$3,816$

    $\dfrac{\$3,816}{9} = \$424/\text{month}$

(c) <u>Rule of 78s</u>:   5 months early out of 9:

Numerator:

$$\dfrac{5(5+1)}{2} = \dfrac{30}{2} = 15$$

Denominator:

$$\dfrac{9(9+1)}{2} = \dfrac{90}{2} = 45$$

$$\dfrac{15}{45} \times \$216 = \$72 \text{ Interest saved}$$

9. (Continued)

 (d) $2,120 Total of 5 remaining payments of $424 each
     -   72 Interest saved
     $2,048 Balance due

11. (a) I = Prt   (b) M = P + I
                                             = $4,600 + $517.50
        = $4,600 × 0.09 × $\frac{15}{12}$ M = $5,117.50

        I = $517.50 $\frac{\$5,117.50}{15}$ = $341.17/month

 (c) <u>Rule of 78s</u>: 9 months early out of 15:

    Numerator: Denominator:

    $\frac{9(9+1)}{2}$ = $\frac{90}{2}$ = 45 $\frac{15(15+1)}{2}$ = $\frac{240}{2}$ = 120

    $\frac{45}{120}$ × $517.50 = $194.06 Interest saved

 (d) $3,070.53 Total of remaining 9 payments of $341.17 each
     - 194.06 Interest saved
     $2,876.47 Balance due

Section 1

1. (a) 6 years × 1 period per year = 6 periods

$$\frac{7\%}{1 \text{ period per year}} = 7\% \text{ per period}$$

(b) 2 years × 4 periods per year = 8 periods

$$\frac{9\%}{4 \text{ periods per year}} = 2.25\% \text{ per period}$$

(c) 4 years × 2 periods per year = 8 periods

$$\frac{10\%}{2 \text{ periods per year}} = 5\% \text{ per period}$$

(d) 3 years × 12 periods per year = 36 periods

$$\frac{6\%}{12 \text{ periods per year}} = 0.5\% \text{ per period}$$

(e) 9 years × 2 periods a year = 18 periods

$$\frac{11\%}{2 \text{ periods per year}} = 5.5\% \text{ per period}$$

3. (a) 4 years × 1 period per year = 4 periods

$$\frac{7\%}{1 \text{ period per year}} = 7\% \text{ per period}$$

Period 1	Principal	$4,000.00	
	Interest	280.00	(7% of $4,000)
Period 2	Principal	$4,280.00	
	Interest	299.60	(7% of $4,280)
Period 3	Principal	$4,579.60	
	Interest	320.57	(7% of $4,579.60)
Period 4	Principal	$4,900.17	
	Interest	343.01	(7% of $4,900.17)
	Compound amount	$5,243.18	

Compound amount $5,243.18
Original principal   -4,000.00
Compound interest $1,243.18

3. (Continued)

   (b) 4 years × 2 periods per year = 8 periods

$$\frac{7\%}{2 \text{ periods per year}} = 3.5\% \text{ per period}$$

Period 1	Principal	$4,000.00	
	Interest	140.00	(3.5% of $4,000)
Period 2	Principal	$4,140.00	
	Interest	144.90	(3.5% of $4,140)
Period 3	Principal	$4,284.90	
	Interest	149.97	(3.5% of $4,284.90)
Period 4	Principal	$4,434.87	
	Interest	155.22	(3.5% of $4,434.87)
Period 5	Principal	$4,590.09	
	Interest	160.65	(3.5% of $4,590.09)
Period 6	Principal	$4,750.74	
	Interest	166.28	(3.5% of $4,750.74)
Period 7	Principal	$4,917.02	
	Interest	172.10	(3.5% of $4,917.02)
Period 8	Principal	$5,089.12	
	Interest	178.12	(3.5% of $5,089.12)
	Compound amount	$5,267.24	

```
Compound amount   $5,267.24
Original principal -4,000.00
Compound interest  $1,267.24
```

5. (a) 1 year × 2 periods a year = 2 periods

$$\frac{5\%}{2 \text{ periods per year}} = 2.5\% \text{ per period}$$

Period 1	Principal	$71,000.00	
	Interest	1,775.00	(2.5% of $71,000)
Period 2	Principal	$72,775.00	
	Interest	1,819.38	(2.5% of $72,775)
	Compound amount	$74,594.38	

```
Compound amount   $74,594.38
Original principal -71,000.00
Compound interest  $ 3,594.38
```

   (b) 1 year × 4 periods a year = 4 periods

$$\frac{5\%}{4 \text{ periods per year}} = 1.25\% \text{ per period}$$

5. (b) (Continued)

Period 1	Principal	$71,000.00	
	Interest	887.50	(1.25% of $71,000)
Period 2	Principal	$71,887.50	
	Interest	898.59	(1.25% of $71,887.50)
Period 3	Principal	$72,786.09	
	Interest	909.83	(1.25% of $72,786.09)
Period 4	Principal	$73,695.92	
	Interest	921.20	(1.25% of $73,695.92)
	Compound amount	$74,617.12	

```
Compound amount   $74,617.12
Original principal -71,000.00
Compound interest $ 3,617.12
```

7. (a) $\dfrac{9}{12} = \dfrac{3}{4}$ year × 4 periods a year = 3 periods

$$\frac{6\%}{4 \text{ periods a year}} = 1.5\% \text{ per period}$$

Period 1	Principal	$7,000.00	
	Interest	105.00	(1.5% of $7,000)
Period 2	Principal	$7,105.00	
	Interest	106.58	(1.5% of $7,105)
Period 3	Principal	$7,211.58	
	Interest	108.17	(1.5% of $7,211.58)
	Compound amount	$7,319.75	

```
Compound amount   $7,319.75
Original principal -7,000.00
Compound interest $  319.75
```

(b) $\dfrac{3}{4}$ year × 12 periods a year = 9 periods

$$\frac{6\%}{12 \text{ periods a year}} = 0.5\% \text{ per period}$$

Period 1	Principal	$7,000.00	
	Interest	35.00	(0.5% of $7,000)
Period 2	Principal	$7,035.00	
	Interest	35.18	(0.5% of $7,035)
Period 3	Principal	$7,070.18	
	Interest	35.35	(0.5% of $7,070.18)

7. (b) (Continued)

Period 4	Principal	$7,105.53	
	Interest	35.53	(0.5% of $7,105.53)
Period 5	Principal	$7,141.06	
	Interest	35.71	(0.5% of $7,141.06)
Period 6	Principal	$7,176.77	
	Interest	35.88	(0.5% of $7,176.77)
Period 7	Principal	$7,212.65	
	Interest	36.06	(0.5% of $7,212.65)
Period 8	Principal	$7,248.71	
	Interest	36.24	(0.5% of $7,248.71)
Period 9	Principal	$7,284.95	
	Interest	36.42	(0.5% of $7,284.95)
	Compound amount	$7,321.37	

```
Compound amount    $7,321.37
Original principal -7,000.00
Compound interest  $  321.37
```

9. More interest is earned when interest is compounded more often.

Section 2

1. (a) $P = \$4,000$ $\qquad$ $M = P(1 + i)^n$
   $i = 7\%$ $\qquad\qquad\qquad$ $= \$4,000(1 + 7\%)^4$
   $n = 4$ $\qquad\qquad\qquad\quad$ $= 4,000(1.3107960)$
   $\qquad\qquad\qquad\qquad\quad$ $M = \$5,243.18$

   $I = M - P$
   $\quad = \$5,243.18 - \$4,000.00$
   $I = \$1,243.18$

   (b) $P = \$4,000$ $\qquad$ $M = P(1 + i)^n$
   $i = 3.5\%$ $\qquad\qquad\qquad$ $= \$4,000(1 + 3.5\%)^8$
   $n = 8$ $\qquad\qquad\qquad\quad$ $= 4,000(1.3168090)$
   $\qquad\qquad\qquad\qquad\quad$ $M = \$5,267.24$

   $I = M - P$
   $\quad = \$5,267.24 - \$4,000.00$
   $I = \$1,267.24$

3. (a) P = $71,000 $M = P(1 + i)^n$
     i = 2.5% $= \$71,000(1 + 2.5\%)^2$
     n = 2   = 71,000(1.050625)
                           M = $74,594.38

     I = M − P
       = $74,594.38 − $71,000.00
     I = $3,594.38

(b) P = $71,000 $M = P(1 + i)^n$
     i = 1.25% $= \$71,000(1 + 1.253\%)^4$
     n = 4   = 71,000(1.050945)
                           M = $74,617.12

     I = M − P
       = $74,617.12 − $71,000.00
     I = $3,617.12

5. (a) P = $7,000 $M = P(1 + i)^n$
     i = 1.5% $= \$7,000(1 + 1.5\%)^3$
     n = 3   = 7,000(1.045678)
                           M = $7,319.75

     I = M − P
       = $7,319.75 − $7,000.00
     I = $319.75

(b) P = $7,000 $M = P(1 + i)^n$
     i = 0.5% $= \$7,000(1 + 0.5\%)^9$
     n = 9   = 7,000(1.045911)
                           M = $7,321.37

     I = M − P
       = $7,321.37 − $7,000.00
     I = $321.37

7. (a) P = $85,000 $M = P(1 + i)^n$

     $i = \dfrac{5}{12}\%$ $= \$85,000\left(1 + \dfrac{5}{12}\%\right)^{36}$

     n = 36   = 85,000(1.1614722)

                           M = $98,725.14

     I = M − P
       = $98,725.14 − $85,000.00
     I = $13,725.14

7. (Continued)

    (b) P = \$1,700 $M = P(1 + i)^n$
        i = 1.5% $= \$1,700(1 + 1.5\%)^8$
        n = 8 $= 1,700(1.126492)$
                                   $M = \$1,915.04$

        I = M - P
          = \$1,915.04 - \$1,700.00
        I = \$215.04

    (c) P = \$45,000 $M = P(1 + i)^n$
        i = 3.5% $= \$45,000(1 + 3.5\%)^{12}$
        n = 12 $= 45,000(1.5110687)$
                                     $M = \$67,998.09$

        I = M - P
          = \$67,998.09 - \$45,000.00
        I = \$22,998.09

    (d) P = \$4,500 $M = P(1 + i)^n$
        $i = \dfrac{5}{6}\%$ $= \$4,500\left(1 + \dfrac{5}{6}\%\right)^{48}$
        n = 48 $= 4,500(1.489354)$
                                     $M = \$6,702.09$

        I = M - P
          = \$6,702.09 - \$4,500.00
        I = \$2,202.09

    (e) P = \$8,100 $M = P(1 + i)^n$
        i = 9% $= \$8,100(1 + 9\%)^7$
        n = 7 $= 8,100(1.8280391)$
                                     $M = \$14,807.12$

        I = M - P
          = \$14,807.12 - \$8,100.00
        I = \$6,707.12

9. (a) P = \$500 $M = P(1 + i)^n$
        $i = \dfrac{5}{12}\%$ $= \$500\left(1 + \dfrac{5}{12}\%\right)^{36}$
        n = 36 $= 500(1.161472)$
                                     $M = \$580.74$

9. (a) (Continued)

$$I = M - P$$
$$ = \$580.74 - \$500.00$$
$$I = \$80.74$$

(b) $P = \$1,000$

$i = \dfrac{5}{12}\%$

$n = 36$

$M = P(1 + i)^n$

$ = \$1,000\left(1 + \dfrac{5}{12}\%\right)^{36}$

$ = 1,000(1.161472)$

$M = \$1,161.47$

$$I = M - P$$
$$ = \$1,161.47 - \$1,000.00$$
$$I = \$161.47$$

(c) $P = \$2,000$

$i = \dfrac{5}{12}\%$

$n = 36$

$M = P(1 + i)^n$

$ = \$2,000\left(1 + \dfrac{5}{12}\%\right)^{36}$

$ = 2,000(1.161472)$

$M = \$2,322.94$

$$I = M - P$$
$$ = \$2,322.94 - \$2,000.00$$
$$I = \$322.94$$

(d) Doubling the principal doubles the interest when interest rate and time are the same.

11. (a) $P = \$1,000$
$i = 1\%$
$n = 8$

$M = P(1 + i)^n$
$ = \$1,000(1 + 1\%)^8$
$ = 1,000(1.082857)$
$M = \$1,082.86$

$$I = M - P$$
$$ = \$1,082.86 - \$1,000.00$$
$$I = \$82.86$$

(b) $P = \$1,000$
$i = 2\%$
$n = 8$

$M = P(1 + i)^n$
$ = \$1,000(1 + 2\%)^8$
$ = 1,000(1.171659)$
$M = \$1,171.66$

$$I = M - P$$
$$ = \$1,171.66 - \$1,000.00$$
$$I = \$171.66$$

11. (Continued)

    (c) P = \$1,000

          i = 4%

          n = 8

$$M = P(1 + i)^n$$
$$= \$1,000(1 + 4\%)^8$$
$$= 1,000(1.368569)$$
$$M = \$1,368.57$$

$$I = M - P$$
$$= \$1,368.57 - \$1,000.00$$
$$I = \$368.57$$

    (d) No

13. (a) P = \$1,000

          i = 3%

          n = 4

$$M = P(1 + i)^n$$
$$= \$1,000(1 + 3\%)^4$$
$$= 1,000(1.125509)$$
$$M = \$1,125.51$$

$$I = M - P$$
$$= \$1,125.51 - \$1,000.00$$
$$I = \$125.51$$

    (b) P = \$1,000

          i = 3%

          n = 8

$$M = P(1 + i)^n$$
$$= \$1,000(1 + 3\%)^8$$
$$= 1,000(1.266770)$$
$$M = \$1,266.77$$

$$I = M - P$$
$$= \$1,266.77 - \$1,000.00$$
$$I = \$266.77$$

    (c) P = \$1,000

          i = 3%

          n = 16

$$M = P(1 + i)^n$$
$$= \$1,000(1 + 3\%)^{16}$$
$$= 1,000(1.604706)$$
$$M = \$1,604.71$$

$$I = M - P$$
$$= \$1,604.71 - \$1,000.00$$
$$I = \$604.71$$

    (d) No

15. $P_1 = \$1,000$ $\qquad$ $P_2 = \$1,500$

$\quad$ (a) $P_1 = \$1,000$ $\qquad$ $M_1 = P_1(1 + i)^n$

$\qquad$ $i = \dfrac{3}{4}\%$ $\qquad\qquad\qquad$ $= \$1,000(1 + \dfrac{3}{4}\%)^2$

$\qquad$ $n = 2$ $\qquad\qquad\qquad\qquad$ $= 1,000(1.015056)$

$\qquad\qquad\qquad\qquad\qquad\qquad\qquad$ $M_1 = \$1,015.06$

$\quad$ (b) $\$1,500.00 - \$1,015.06 = \$484.94$

$\quad$ (c) $P_2 = \$1,500$ $\qquad$ $M_2 = P_2(1 + i)^n$

$\qquad$ $i = \dfrac{3}{4}\%$ $\qquad\qquad\qquad$ $= \$1,500(1 + \dfrac{3}{4}\%)^4$

$\qquad$ $n = 4$ $\qquad\qquad\qquad\qquad$ $= 1,500(1.030339)$

$\qquad\qquad\qquad\qquad\qquad\qquad\qquad$ $M_2 = \$1,545.51$

$\quad$ (d)

$\$1,545.51$	Maturity$_2$	$I_1 =$	$\$15.06$
$-1,484.94$	Total deposits (or)	$I_2 =$	$+45.51$
$\$\quad 60.57$			$\$60.57$

17. $P_1 = \$1,000$ $\qquad$ $P_2 = \$1,400$

$\quad$ (a) $P_1 = \$1,000$ $\qquad$ $M_1 = P_1(1 + i)^n$

$\qquad$ $i = \dfrac{5}{12}\%$ $\qquad\qquad\qquad$ $= \$1,000\left(1 + \dfrac{5}{12}\%\right)^3$

$\qquad$ $n = 3$ $\qquad\qquad\qquad\qquad$ $= 1,000(1.012552)$

$\qquad\qquad\qquad\qquad\qquad\qquad\qquad$ $M_2 = \$1,012.55$

$\quad$ (b) $\$1,400.00 - \$1,012.55 = \$387.45$

$\quad$ (c) $P_2 = \$1,400$ $\qquad$ $M_2 = P_2(1 + i)^n$

$\qquad$ $i = 1.5\%$ $\qquad\qquad\qquad$ $= \$1,400(1 + 1.5\%)^4$

$\qquad$ $n = 4$ $\qquad\qquad\qquad\qquad$ $= 1,400(1.061364)$

$\qquad\qquad\qquad\qquad\qquad\qquad\qquad$ $M_2 = \$1,485.91$

$\quad$ (d)

Interest, CD#1	$\$12.55$	$\$1,485.91$	Maturity$_2$
Interest, CD#2	$+85.91$ (or)	$-1,387.45$	Total deposits
Total interest	$\$98.46$	$\$\quad 98.46$	Total interest

19. $P_1 = \$1,000$ $P_2 = \$1,400$

    (a) $P_1 = \$1,000$ $M_1 = P_1(1 + i)^n$

         $i = \dfrac{5}{12}\%$ $= \$1,000\left(1 + \dfrac{5}{12}\%\right)^6$

         $n = 6$ $= 1,000(1.025262)$

                                $M_1 = \$1,025.26$

    (b) $\$1,400.00 - \$1,025.26 = \$374.74$

    (c) $P_2 = \$1,400$ $M_2 = P_2(1 + i)^n$
         $i = 1.75\%$ $= \$1,400(1 + 1.75\%)^8$
         $n = 8$ $= 1,400(1.148882)$
                                $M_2 = \$1,608.43$

    (d) Interest, CD#1 $\$ 25.26$ $\$1,608.43$ Maturity$_2$
         Interest, CD#2 $\underline{+208.43}$ (or) $\underline{-1,374.74}$ Total deposits
         Total interest $\$233.69$ $\$ 233.69$ Total interest

Section 3

1. (a) $P = \$9,300$

         $I = P \times$ Dep. tab. $M = P + I$
           $= \$9,300(0.0072759)$ $= \$9,300.00 + \$67.67$
         $I = \$67.67$ $M = \$9,367.67$

    (b) $P = \$800$

         $I = P \times$ Dep. tab. $M = P + I$
           $= \$800(0.01068)$ $= \$800.00 + \$8.54$
         $I = \$8.54$ $M = \$808.54$

    (c) $P = \$4,000$

         $I = P \times$ Dep. tab. $M = P + I$
           $= \$4,000(0.003381)$ $= \$4,000.00 + \$13.52$
         $I = \$13.52$ $M = \$4,013.52$

    (d) $P = \$600$

         $I = P \times$ Dep. tab. $M = P + I$
           $= \$600(0.01043)$ $= \$600.00 + \$6.26$
         $I = \$6.26$ $M = \$606.26$

3. (a-1) P = $12,000

        I = P × Dep. tab.
          = $12,000(0.0113128)
        I = $135.75

(a-2) P = $12,000

        I = P × Dep. tab.
          = $12,000(0.01125)
        I = $135.00

(b-1) P = $4,500

        I = P × Dep. tab.
          = $4,500(0.0113128)
        I = $50.91

(b-2) P = $4,500

        I = P × Dep. tab.
          = $4,500(0.01125)
        I = $50.63

5. (a) I = P × Dep. tab.

			Balance
$I_1 = \$700(0.0113128)$	$= \$ 7.92$		$\$ 700.00$
$I_2 = \$500(0.0086618)$	$= 4.33$		500.00
$I_3 = \$300(0.0057662)$	$= 1.73$		300.00
$I_4 = \$400(0.0032551)$	$= \underline{1.30}$		400.00
Total interest	$\$15.28$		$\underline{15.28}$
			$\$1,915.28$

(b) I = P × Dep. tab.

			Balance
$I_1 = \$4,300(0.0113128)$	$= \$48.65$		$\$4,300.00$
$I_2 = \$ 200(0.0076537)$	$= 1.53$		200.00
$I_3 = \$ 400(0.0060177)$	$= 2.41$		400.00
$I_4 = \$ 300(0.0027536)$	$= \underline{0.83}$		300.00
Total interest	$\$ 53.42$		$\underline{53.42}$
			$\$5,253.42$

7. (a) P = $10,000 - $1,000 = $9,000

		Interest	Balance
$I_1 = $ P × Dep. tab.		$\$101.82$	$\$10,000.00$
$= \$9,000(0.0113128)$		$\underline{4.26}$	$- 1,000.00$
$I_1 = \$101.82$		$\$106.08$	$\underline{+ \quad 106.08}$
			$\$ 9,106.08$
$I_2 = $ W × W/D tab.			
$= \$1,000(0.004259)$			
$I_2 = \$4.26$			

(b) P = $5,500 - $100 - $400 = $5,000

		Interest	Balance
$I_1 = $ P × Dep. tab.		$\$56.56$	$\$5,500.00$
$= \$5,000(0.0113128)$		0.50	$- 100.00$
$I_1 = \$56.56$		$\underline{3.06}$	$- 400.00$
		$\$60.12$	$\underline{+ \quad 60.12}$
$I_2 = $ W × W/D tab.			$\$5,060.12$
$= \$100(0.00501)$			
$I_2 = \$0.50$			

7. (b) (Continued)

$$I_3 = W \times W/D \text{ tab.}$$
$$= \$400(0.00765)$$
$$I_3 = \$3.06$$

9. (a) $P_1 = \$13,400 - \$1,000 = \$12,400$
$P_2 = \$4,100$

$I_1 = P_1 \times$ Dep. tab.	Interest	Balance
$= \$12,400(0.0113128)$	$140.28	$13,400.00
$I_1 = \$140.28$	31.38	-1,000.00
	5.14	+4,100.00
$I_2 = P_2 \times$ Dep. tab.	$176.80	+ 176.80
$= \$4,100(0.007654)$		$16,676.80
$I_2 = \$31.38$		

$$I_3 = W \times W/D \text{ tab.}$$
$$= \$1,000(0.005138)$$
$$I_3 = \$5.14$$

(b) $P_1 = \$4,900 - \$600 - \$400 = \$3,900$
$P_2 = \$800$
$P_3 = \$1,000$

$I_1 = P_1 \times$ Dep. tab.	Interest	Balance
$= \$3,900(0.0113128)$	$44.12	$4,900.00
$I_1 = \$44.12$	5.62	- 600.00
	2.38	- 400.00
$I_2 = P_2 \times$ Dep. tab.	1.13	+ 800.00
$= \$800(0.00702)$	4.02	+1,000.00
$I_2 = \$5.62$	$57.27	+ 57.27
		$5,757.27

$$I_3 = P_3 \times \text{Dep. tab.}$$
$$= \$1,000(0.002378)$$
$$I_3 = \$2.38$$

$$I_4 = W \times W/D \text{ tab.}$$
$$= \$600(.00188)$$
$$I_4 = \$1.13$$

$$I_5 = W \times W/D \text{ tab.}$$
$$= \$400(.01005)$$
$$I_5 = \$4.02$$

(c) $P_1 = \$5,100 - \$200 - \$900 = \$4,000$
$P_2 = \$400$
$P_3 = \$100$

9. (c) (Continued)

	Interest	Balance
$I_1 = P_1 \times$ Dep. tab.	$45.25	$5,100.00
$= \$4,000(0.0113128)$	4.17	$-$ 200.00
$I_1 = \$45.25$	0.65	$-$ 900.00
	0.68	$+$ 400.00
$I_2 = P_2 \times$ Dep. tab.	9.04	$+$ 100.00
$= \$400(0.0104284)$	$\overline{\$59.79}$	$+$ 59.79
$I_2 = \$4.17$		$\overline{\$4,559.79}$

$I_3 = P_3 \times$ Dep. tab.
$\quad = \$100(0.0065208)$
$I_3 = \$0.49$

$I_4 = W \times$ W/D tab.
$\quad = \$200(0.0033805)$
$I_4 = \$0.68$

$I_5 = W \times$ W/D tab.
$\quad = \$900(0.0100495)$
$I_5 = \$9.04$

Section 4

1. (a) $M = \$14,600$ $P = M(1 + i)^{-n}$
     $i = 2\%$ $= \$14,600(1 + 2\%)^{-4}$
     $n = 4$ $= 14,600(0.9238454)$
                               $P = \$13,488.14$

     $I = M - P$
       $= \$14,600.00 - \$13,488.14$
     $I = \$1,111.86$

(b) $M = \$2,500$ $P = M(1 + i)^{-n}$
     $i = 1.25\%$ $= \$2,500(1 + 1.25\%)^{-20}$
     $n = 20$ $= 2,500(0.780009)$
                               $P = \$1,950.02$

     $I = M - P$
       $= \$2,500.00 - \$1,950.02$
     $I = \$549.98$

(c) $M = \$6,800$ $P = M(1 + i)^{-n}$
     $i = \dfrac{7}{12}\%$ $= \$6,800\left(1 + \dfrac{7}{12}\%\right)^{-48}$
     $n = 48$ $= 6,800(0.756399)$

                               $P = \$5,143.51$

1. (c) (Continued)

$$I = M - P$$
$$= \$6,800.00 - \$5,143.51$$
$$I = \$1,656.49$$

(d) $M = \$9,500$ $P = M(1 + i)^{-n}$
     $i = 2.5\%$ $= \$9,500(1 + 2.5\%)^{-2}$
     $n = 2$ $= 9,500(0.951814)$
                         $P = \$9,042.23$

$$I = M - P$$
$$= \$9,500.00 - \$9,042.23$$
$$I = \$457.77$$

(e) $M = \$4,400$ $P = M(1 + i)^{-n}$
     $i = 1.5\%$ $= \$4,400(1 + 1.5\%)^{-12}$
     $n = 12$ $= 4,400(0.836387)$
                         $P = \$3,680.10$

$$I = M - P$$
$$= \$4,400.00 - \$3,680.10$$
$$I = \$719.90$$

3. (a) $M = \$5,000$ $P = M(1 + i)^{-n}$
     $i = 0.5\%$ $= \$5,000(1 + 0.5\%)^{-12}$
     $n = 12$ $= 5,000(0.941905)$
                         $P = \$4,709.53$

(b) $I = M - P$
     $= \$5,000.00 - \$4,709.53$
     $I = \$290.47$

5. (a) $M = \$225,000$ $P = M(1 + i)^{-n}$
     $i = 2\%$ $= \$225,000(1 + 2\%)^{-8}$
     $n = 8$ $= 225,000(0.8534904)$
                         $P = \$192,035.34$

(b) $I = M - P$
     $= \$225,000.00 - \$192,035.34$
     $I = \$32,964.66$

7. (a) $M = \$100,000$ $P = M(1 + i)^{-n}$
     $i = 4.5\%$ $= \$100,000(1 + 4.5\%)^{-12}$
     $n = 12$ $= 100,000(0.5896639)$
                         $P = \$58,966.39$

7. (Continued)

    (b) $I = M - P$
          $= \$100,000.00 - \$58,966.39$
       $I = \$41,033.61$

9. (a) $P = \$85,000$ $M = P(1 + i)^n$
       $i = 1.5\%$ $= \$85,000(1 + 1.5\%)^{16}$
       $n = 16$ $= 85,000(1.2689855)$
                           $M = \$107,863.76$

    (b) $M = \$107,863.76$ $P = M(1 + i)^{-n}$

$$i = \frac{5}{12}\% \qquad = \$107,863.76\left(1 + \frac{5}{12}\%\right)^{-48}$$

       $n = 48$ $= 107,863.76(0.8190710)$

                        $P = \$88,348.08$

11. (a) $P = \$6,000$ $I = Prt$ $M = P + I$
        $r = 11\%$ $= \$6,000 \times 0.11 \times 2$ $= \$6,000 + \$1,320$
        $t = 2$ yrs. $I = \$1,320$ $M = \$7,320$

    (b) $M = \$7,320$ $P = M(1 + i)^{-n}$

$$i = \frac{7}{12}\% \qquad = \$7,320\left(1 + \frac{7}{12}\%\right)^{-24}$$

       $n = 24$ $= 7,320(0.869712)$

                    $P = \$6,366.29$

13. (a) $P = \$40,000$ $I = Prt$ $M = P + I$
        $r = 13.5\%$ $= \$40,000 \times 0.135 \times 1$ $= \$40,000 + \$5,400$
        $t = 1$ yr. $I = \$5,400$ $M = \$45,400$

    (b) $M = \$45,400$ $P = M(1 + i)^{-n}$
        $i = 0.5\%$ $= \$45,400(1 + 0.5\%)^{-12}$
        $n = 12$ $= 45,400(0.9419053)$
                  $P = \$42,762.50$

Section 1

1. (a) Pmt. = $2,200 $M = \text{Pmt.} \times \text{Amt. ann. tab.}_{\overline{n}|i}$

      $i = 6\%$ $= \$2,200 \times \text{Amt. ann. tab.}_{\overline{20}|6\%}$

      $n = 20$ $= 2,200(36.785591)$

                        $M = \$80,928.30$

    Total deposits = $\$2,200 \times 20$ payments = $44,000
    Total interest = $80,928.30 - $44,000.00 = $36,928.30

   (b) Pmt. = $800 $M = \text{Pmt.} \times \text{Amt. ann. tab.}_{\overline{n}|i}$

      $i = 1.75\%$ $= \$800 \times \text{Amt. ann. tab.}_{\overline{12}|1.75\%}$

      $n = 12$ $= 800(13.225104)$

                        $M = \$10,580.08$

    Total deposits = $\$800 \times 12$ payments = $9,600
    Total interest = $10,580.08 - $9,600.00 = $980.08

   (c) Pmt. = $3,000 $M = \text{Pmt.} \times \text{Amt. ann. tab.}_{\overline{n}|i}$

      $i = 4\%$ $= \$3,000 \times \text{Amt. ann. tab.}_{\overline{16}|4\%}$

      $n = 16$ $= 3,000(21.824531)$

                        $M = \$65,473.59$

    Total deposits = $\$3,000 \times 16$ payments = $48,000
    Total interest = $65,473.59 - $48,000.00 = $17,473.59

   (d) Pmt. = $1,500 $M = \text{Pmt.} \times \text{Amt. ann. tab.}_{\overline{n}|i}$

      $i = 1.25\%$ $= \$1,500 \times \text{Amt. ann. tab.}_{\overline{24}|1.25\%}$

      $n = 24$ $= 1,500(27.788084)$

                        $M = \$41,682.13$

    Total deposits = $\$1,500 \times 24$ payments = $36,000
    Total interest = $41,682.13 - $36,000.00 = $5,682.13

   (e) Pmt. = $1,100 $M = \text{Pmt.} \times \text{Amt. ann. tab.}_{\overline{n}|i}$

      $i = 0.5\%$ $= \$1,100 \times \text{Amt. ann. tab.}_{\overline{36}|0.5\%}$

      $n = 36$ $= 1,100(39.336105)$

                        $M = \$43,269.72$

    Total deposits = $\$1,100 \times 36$ payments = $39,600
    Total interest = $43,269.72 - $39,600.00 = $3,669.72

3. Pmt. = $500   (a) $M = $ Pmt. $\times$ Amt. ann. tab.$_{n|i}$

    i = 1.75% $= \$500 \times$ Amt. ann. tab.$_{8|1.75\%}$

    n = 8 $= 500(8.507530)$

                        $M = \$4,253.77$

(b) Total deposits = $500 × 8 = $4,000

(c) Total interest = $4,253.77 - $4,000.00 = $253.77

5. Pmt. = $500   (a) $M = $ Pmt. $\times$ Amt. ann. tab.$_{n|i}$

    i = 0.5% $= \$500 \times$ Amt. ann. tab.$_{60|0.5\%}$

    n = 60 $= 500(69.770031)$

                        $M = \$34,885.02$

(b) Total deposits = $500 × 60 = $30,000

(c) Total interest = $34,885.02 - $30,000.00 = $4,885.02

7. (a) Total investment = $6,000 × 5 payments = $30,000

(b) Pmt. = $6,000 $M = $ Pmt. $\times$ Amt. ann. tab.$_{n|i}$

    i = 8% $= \$6,000 \times$ Amt. ann. tab.$_{5|8\%}$

    n = 5 $= 6,000(5.8666010)$

                      $M = \$35,199.61$

(c) Total interest = $35,199.61 - $30,000.00 = $5,199.61

9. (a) Total investment = $8,000 × 40 = $320,000

(b) Pmt. = $8,000 $M = $ Pmt. $\times$ Amt. ann. tab.$_{n|i}$

    i = 1.25% $= \$8,000 \times$ Amt. ann. tab.$_{40|1.25\%}$

    n = 40 $= 8,000(51.489557)$

                      $M = \$411,916.45$

(c) Total interest = $411,916.45 - $320,000.00 = $91,916.45

Section 2

1. (a) (1) Pmt. = \$2,500   P.V. = Pmt. $\times$ P.V. ann. tab.$_{\overline{n}|i}$

              i = 7.5%   = \$2,500 $\times$ P.V. ann. tab.$_{\overline{16}|7.5\%}$

              n = 16   = 2,500(9.1415067)

                        P.V. = \$22,853.77

    (2) \$2,500 $\times$ 16 = \$40,000

    (3) \$40,000.00 - \$22,853.77 = \$17,146.23

  (b) (1) Pmt. = \$1,500   P.V. = Pmt. $\times$ P.V. ann. tab.$_{\overline{n}|i}$

              i = 0.75%   = \$1,500 $\times$ P.V. ann. tab.$_{\overline{36}|0.75\%}$

              n = 36   = 1,500(31.446805)

                        P.V. = \$47,170.21

    (2) \$1,500 $\times$ 36 = \$54,000

    (3) \$54,000.00 - \$47,170.21 = \$6,829.79

  (c) (1) Pmt. = \$3,000   P.V. = Pmt. $\times$ P.V. ann. tab.$_{\overline{n}|i}$

              i = 2%   = \$3,000 $\times$ P.V. ann. tab.$_{\overline{40}|2\%}$

              n = 40   = 3,000(27.355479)

                        P.V. = \$82,066.44

    (2) \$3,000 $\times$ 40 = \$120,000

    (3) \$120,000.00 - \$82,066.44 = \$37,933.56

  (d) (1) Pmt. = \$1,700   P.V. = Pmt. $\times$ P.V. ann. tab.$_{\overline{n}|i}$

              i = 3%   = \$1,700 $\times$ P.V. ann. tab.$_{\overline{24}|3\%}$

              n = 24   = 1,700(16.935542)

                        P.V. = \$28,790.42

    (2) \$1,700 $\times$ 24 = \$40,800

    (3) \$40,800.00 - \$28,790.42 = \$12,009.58

  (e) (1) Pmt. = \$2,600   P.V. = Pmt. $\times$ P.V. ann. tab.$_{\overline{n}|i}$

              i = 3%   = \$2,600 $\times$ P.V. ann. tab.$_{\overline{10}|3\%}$

              n = 10   = 2,600(8.530203)

                      P.V. = \$22,178.53

1. (e) (Continued)

    (2) $2,600 × 10 = $26,000

    (3) $26,000.00 - $22,178.53 = $3,821.47

3. (a) (1) Pmt. = $6,000   M = Pmt. × Amt. ann. tab.$_{\overline{n}|i}$
         i = 2%   = $6,000 × Amt. ann. tab.$_{\overline{32}|2\%}$
         n = 32   = 6,000(44.227030)
                     M = $265,362.18

    Total payments = $6,000 × 32 = $192,000
    Interest = $265,362.18 - $192,000.00 = $73,362.18

    (2) P.V. = Pmt. × P.V. ann. tab.$_{\overline{n}|i}$
        = $6,000 × P.V. ann. tab.$_{\overline{32}|2\%}$
        = 6,000(23.468335)
    P.V. = $140,810.01

    Interest = $192,000.00 - $140,810.01 = $51,189.99

(b) (1) Pmt. = $300   M = Pmt. × Amt. ann. tab.$_{\overline{n}|i}$
         i = 3.5%   = $300 × Amt. ann. tab.$_{\overline{24}|3.5\%}$
         n = 24   = 300(36.66653)
                     M = $10,999.96

    Total payments = $300 × 24 = $7,200
    Interest = $10,999.96 - $7,200.00 = $3,799.96

    (2) P.V. = Pmt. × P.V. ann. tab.$_{\overline{n}|i}$
        = $300 × P.V. ann. tab.$_{\overline{24}|3.5\%}$
        = 300(16.05837)
    P.V. = $4,817.51

    Interest = $7,200.00 - $4,817.51 = $2,382.49

(c) (1) Pmt. = $4,900   M = Pmt. × Amt. ann. tab.$_{\overline{n}|i}$
         i = 0.5%   = $4,900 × Amt. ann. tab.$_{\overline{48}|0.5\%}$
         n = 48   = 4,900(54.097832)
                     M = $265,079.37

    Total payments = $4,900 × 48 = $235,200
    Interest = $265,079.37 - $235,200.00 = $29,879.37

3. (c) (Continued)

(2) P.V. = Pmt. × P.V. ann. tab.$_{\overline{n}|i}$

P.V. = \$4,900 × P.V. ann. tab.$_{\overline{48}|0.5\%}$

= 4,900(42.580318)

P.V. = \$208,643.55

Interest = \$235,200.00 − \$208,643.55 = \$26,556.45

5. Pmt. = \$1,700   (a) P.V. = Pmt. × P.V. ann. tab.$_{\overline{n}|i}$

i = 1.5%   = \$1,700 × P.V. ann. tab.$_{\overline{12}|1.5\%}$

n = 12   = 1,700(10.907505)

P.V. = \$18,542.76

(b) \$1,700 × 12 = \$20,400

(c) \$20,400.00 − \$18,542.76 = \$1,857.24

7. Pmt. = \$600   (a) P.V. = Pmt. × P.V. ann. tab.$_{\overline{n}|i}$

i = 1.25%   = \$600 × P.V. ann. tab.$_{\overline{8}|1.25\%}$

n = 8   = 600(7.568124)

P.V. = \$4,540.87

(b) \$600 × 8 = \$4,800

(c) \$4,800.00 − \$4,540.87 = \$259.13

9. Pmt. = \$700   (a) P.V. = Pmt. × P.V. ann. tab.$_{\overline{n}|i}$

i = 4%   = \$700 × P.V. ann. tab.$_{\overline{8}|4\%}$

n = 8   = 700(6.73274)

= \$4,712.92

P.V. = \$4,700 (to nearest hundred)

9. (Continued)

    (b) M = \$4,700 $P = M(1 + i)^{-n}$

         i = 4% $= \$4,700(1 + 4\%)^{-20}$

         n = 20 $= 4,700(0.45638)$

                              $= \$2,145.02$

                          P = \$2,100 (to nearest hundred)

    (c) \$700 × 8 = \$5,600

        \$5,600 - \$2,100 = \$3,500

11. (a) Pmt. = \$800   P.V. = Pmt. × P.V. ann. tab.$_{n|i}$

         i = 0.5%   = \$800 × P.V. ann. tab.$_{48|0.5\%}$

         n = 48   = 800(42.580318)

                      P.V. = \$34,064.25 or \$34,100

    (b) M = \$34,100 $P = M(1 + i)^{-n}$

        i = 0.5% $= \$34,100(1 + 0.5\%)^{-72}$

        n = 72 $= 34,100(0.6983024)$

                     $P = \$23,812.11$ or \$23,800

    (c) \$800 × 48 = \$38,400

        \$38,400 - \$23,800 = \$14,600

13. (a) \$3,000 × 72 = \$216,000 + \$25,000 down payment = \$241,000
        total cost

    (b) Pmt. = \$3,000   P.V. = Pmt. × P.V. ann. tab.$_{n|i}$

         $i = \frac{5}{6}\%$ $= \$3,000 × $ P.V. ann. tab.$_{72|\frac{5}{6}\%}$

         n = 72   = 3,000(53.978665)

                      P.V. = \$161,935.99

13. (b) (Continued)

Equivalent cash price: $161,935.99
                        + 25,000.00 down payment
                        $186,935.99

15. (a) $3,000 × 12 = $36,000 + $3,000 down payment = $39,000
        total cost

    (b) Pmt. = $3,000   P.V. = Pmt. × P.V. ann. tab.$_{n|i}$
         i = 2.25%   = $3,000 × P.V. ann. tab.$_{12|2.25\%}$
         n = 12   = 3,000(10.414779)
                          P.V. = $31,244.34

        Equivalent cash price: $31,244.34
                                + 3,000.00 down payment
                                $34,244.34

Section 3

1. (a) (1) M = $30,000   Pmt. = M × s.f. tab.$_{n|i}$
           i = 1.5%   = $30,000 × s.f. tab.$_{20|1.5\%}$
           n = 20   = 30,000(0.0432457)
                             Pmt. = $1,297.37

       (2) Total payments = $1,297.37 × 20 = $25,947.40

       (3) Total interest = $30,000.00 - $25,947.40 = $4,052.60

   (b) (1) M = $50,000   Pmt. = M × s.f. tab.$_{n|i}$
           i = 4%   = $50,000 × s.f. tab.$_{12|4\%}$
           n = 12   = 50,000(0.0665522)
                             Pmt. = $3,327.61

       (2) Total payments = $3,327.61 × 12 = $39,931.32

       (3) Total interest = $50,000.00 - $39,931.32 = $10,068.68

1. (Continued)

(c) (1) M = $60,000   Pmt. = M × s.f. tab.$_{n|i}$

      i = 6%   = $60,000 × s.f. tab.$_{7|6\%}$

      n = 7   = 60,000(0.119135)

                          Pmt. = $7,148.10

    (2) Total payments = $7,148.10 × 7 = $50,036.70

    (3) Total interest = $60,000.00 − $50,036.70 = $9,963.30

(d) (1) M = $100,000   Pmt. = M × s.f. tab.$_{n|i}$

      i = 1.25%   = $100,000 × s.f. tab.$_{11|1.25\%}$

      n = 16   = 100,000(0.05684672)*

                          Pmt. = $5,684.67

                        *Mental math used since most calculators will not take 8 decimal places.

    (2) Total payments = $5,684.67 × 16 = $90,954.72

    (3) Total interest = $100,000.00 − $90,954.72 = $9,045.28

(e) (1) M = $200,000   Pmt. = M × s.f. tab.$_{n|i}$

      i = 0.75%   = $200,000 × s.f. tab.$_{36|0.75\%}$

      n = 36   = 200,000(0.0242997)

                          Pmt. = $4,859.94

    (2) Total payments = $4,859.94 × 36 = $174,957.84

    (3) Total interest = $200,000.00 − $174,957.84
                              = $25,042.16

3.

	Price Quote	Selling Price	Premium or Discount	Interest Rate	Annual Interest	Current Yield
(a)	105	$1,050.00	$ 50P	$6\frac{1}{4}$%	$ 62.50	5.95%
(b)	88	880.00	120D	$5\frac{1}{8}$	51.25	5.82
(c)	100	1,000.00	Par	$9\frac{3}{4}$	97.50	9.75
(d)	$95\frac{1}{2}$	955.00	45D	$8\frac{1}{2}$	85.00	8.90
(e)	$106\frac{1}{4}$	1,062.50	62.50P	10	100.00	9.41

(a) 105% × $1,000 = $1,050
$1,050 - $1,000 = $50P
0.0625 × $1,000 = $62.50

(b) 88% × $1,000 = $880
$1,000 - $800 = $120D
0.05125 × $1,000 = $51.25

Cur. yield = $\dfrac{\text{Interest}}{\text{Price}}$

$= \dfrac{\$62.50}{\$1,050}$

c.y. = 5.95%

Cur. yield = $\dfrac{\text{Interest}}{\text{Price}}$

$= \dfrac{\$51.25}{\$880}$

c.y. = 5.82%

(c) 100% × $1,000 = $1,000 (Par)
0.0975 × $1,000 = $97.50

(d) 95.5% × $1,000 = $955
$1,000 - $955 = $45D
0.085 × $1,000 = $85

Cur. yield = $\dfrac{\text{Interest}}{\text{Price}}$

$= \dfrac{\$97.50}{\$1,000}$

c.y. = 9.75%

Cur. yield = $\dfrac{\text{Interest}}{\text{Price}}$

$= \dfrac{\$85}{\$955}$

c.y. = 8.9%

3. (Continued)

(e) $106.25\% \times \$1,000 = \$1,062.50$
$\$1,062.50 - \$1,000 = \$62.50P$
$0.10 \times \$1,000 = \$100$

$$\text{Cur. yield} = \frac{\text{Interest}}{\text{Price}}$$

$$= \frac{\$100}{\$1,062.50}$$

$$\text{c.y.} = 9.41\%$$

5. (a) $M = \$800,000$ $\text{Pmt.} = M \times \text{s.f. tab.}_{n|i}$

$i = 2.25\%$ $= \$800,000 \times \text{s.f. tab.}_{16|2.25\%}$

$n = 16$ $= 800,000(0.0526166)$

$\text{Pmt.} = \$42,093.28$

(b) $\$42,093.28 \times 16 = \$673,492.48$

(c) $\$800,000 - \$673,492.48 = \$126,507.52$

7. (a) $M = \$25,000$ $\text{Pmt.} = M \times \text{s.f. tab.}_{n|i}$

$i = 3.5\%$ $= \$25,000 \times \text{s.f. tab.}_{4|3.5\%}$

$n = 4$ $= 25,000(0.2372511)$

$\text{Pmt.} = \$5,931.28$

(b) $\$5,931.28 \times 4 = \$23,725.12$

(c) $\$25,000.00 - \$23,725.12 = \$1,274.88$

9. (a) $M = \$7,300,000$ $\text{Pmt.} = M \times \text{s.f. tab.}_{n|i}$

$i = \dfrac{5}{12}\%$ $= \$7,300,000 \times \text{s.f. tab.}_{96|\frac{5}{12}\%}$

$n = 96$ $= 7,300,000(0.0084933)$

$\text{Pmt.} = \$62,001.09$

(b) $\$62,001.09 \times 96 = \$5,952,104.60$

(c) $\$7,300,000.00 - \$5,952,104.60 = \$1,347,895.40$

11. (a) 104% of $1,000 = $1,040

   (b) $1,040 - $1,000 = $40 premium per bond

   (c) $1,040 × 500 bonds = $520,000

   (d) 8.25% × $1,000 = $82.50 interest per year per bond

   (e) Current yield = $\dfrac{\text{Annual interest}}{\text{Current price}} = \dfrac{\$82.50}{\$1,040} = 7.93\%$

   (f) $82.50 × 500 bonds = $41,250

13. (a) M = $8,000   Pmt. = M × s.f. tab.$_{n|i}$

   i = 2%   = $8,000 × s.f. tab.$_{8|2\%}$

   n = 8   = 8,000(0.116510)

                    Pmt. = $932.08

Payment	Pd. Interest (i = 2%)	Periodic Payment	Total Increase	Balance End of Period
1	0	$ 932.08	$ 932.08	$ 932.08*
2	$ 18.64	932.08	950.72	1,882.80
3	37.66	932.08	969.74	2,852.54
4	57.05	932.08	989.13	3,841.67
5	76.83	932.08	1,008.91	4,850.58
6	97.01	932.08	1,029.09	5,879.67
7	117.59	932.08	1,049.67	6,929.34
8	138.59	932.08	1,070.67	8,000.01
	$543.37	$7,456.64		

*If the Excel spreadsheet is used, there may be differences of
a few cents in the answers. The software stores and saves
the numbers with full precision even though the student
rounded to the nearest cent.

   (b) M = $20,000   Pmt. = M × s.f. tab.$_{n|i}$

   i = 6%   = $20,000 × s.f. tab.$_{5|6\%}$

   n = 5   = 20,000(0.177396)

                    Pmt. = $3,547.92

13. (b) (Continued)

Payment	Pd. Interest (i = 6%)	Periodic Payment	Total Increase	Balance End of Period
1	0	$ 3,547.92	$3,547.92	$ 3,547.92*
2	$ 212.88	3,547.92	3,760.80	7,308.72
3	438.52	3,547.92	3,986.44	11,295.16
4	677.71	3,547.92	4,225.63	15,520.79
5	931.25	3,547.92	4,479.17	19,999.96
	$2,260.36	$17,739.60		

*See note at the end of Problem 13a.

Section 4

1. (a) (1) P.V. = $300,000 $\quad$ Pmt. = P.V. $\times$ Amtz. tab.$_{\overline{n}|i}$

$\qquad$ i = 4% $\qquad\qquad\qquad$ = $300,000 $\times$ Amtz. tab.$_{\overline{30}|4\%}$

$\qquad$ n = 30 $\qquad\qquad\qquad$ = 300,000(0.0578301)

$\qquad\qquad\qquad\qquad\qquad$ Pmt. = $17,349.03

$\quad$ (2) $17,349.03 $\times$ 30 = $520,470.90

$\quad$ (3) $520,470.90 - $300,000.00 = $220,470.90

(b) (1) P.V. = $150,000 $\quad$ Pmt. = P.V. $\times$ Amtz. tab.$_{\overline{n}|i}$

$\qquad$ i = 2.25% $\qquad\qquad\qquad$ = $150,000 $\times$ Amtz. tab.$_{\overline{40}|2.25\%}$

$\qquad$ n = 40 $\qquad\qquad\qquad$ = 150,000(0.0381774)

$\qquad\qquad\qquad\qquad\qquad$ Pmt. = $5,726.61

$\quad$ (2) $5,726.61 $\times$ 40 = $229,064.40

$\quad$ (3) $229,064.40 - $150,000.00 = $79,064.40

(c) (1) P.V. = $80,000 $\quad$ Pmt. = P.V. $\times$ Amtz. tab.$_{\overline{n}|i}$

$\qquad$ i = $\frac{7}{12}$% $\qquad\qquad\qquad$ = $80,000 $\times$ Amtz. tab.$_{\overline{60}|\frac{7}{12}\%}$

$\qquad$ n = 60 $\qquad\qquad\qquad$ = 80,000(0.0198012)

$\qquad\qquad\qquad\qquad\qquad$ Pmt. = $1,584.10

$\quad$ (2) $1,584.10 $\times$ 60 = $95,046

$\quad$ (3) $95,046 - $80,000 = $15,046

1. (Continued)

(d) (1) P.V. = $500,000   Pmt. = P.V. × Amtz. tab.$_{n|i}$

i = 5%   = $500,000 × Amtz. tab.$_{14|5\%}$

n = 14   = 500,000(1.01023970)

Pmt. = $50,511.95

(2) $50,511.95 × 14 = $707,167.30

(3) $707,167.30 - $500,000.00 = $207,167.30

(e) (1) P.V. = $650,000   Pmt. = P.V. × Amtz. tab.$_{n|i}$

i = 9%   = $650,000 × Amtz. tab.$_{20|9\%}$

n = 20   = 650,000(0.1095465)

Pmt. = $71,205.23

(2) $71,205.23 × 20 = $1,424,104.60

(3) $1,424,104.60 - $650,000.00 = $774,104.60

3. (a) $ 7.34 per $1,000 @ 8%   (b) $ 8.84 per $1,000 @ 8.75%
   ×   200   ×   140
   $ 1,468 Monthly payment $1,237.60 Monthly payment
   ×   360 Mos. in 30 yrs.   ×   240 Mos. in 20 yrs.
   $528,480 Total payments $297,024 Total payments
   -200,000 Principal   -140,000 Principal
   $328,480 Total interest $157,024 Total interest

(c) $ 8.57 per $1,000 @ 9.25%   (d) $ 9.00 per $1,000 @ 9%
   ×   125   ×   90
   $1071.25 Monthly payment $ 810 Monthly payment
   ×   300 Mos. in 25 yrs.   ×   240 Mos. in 20 yrs.
   $321,375 Total payments $194,400 Total payments
   -125,000 Principal   - 90,000 Principal
   $196,375 Total interest $104,400 Total interest

5. (a) P.V. = $30,000   Pmt. = P.V. × Amtz. tab.$_{n|i}$

i = 0.75%   = $30,000 × Amtz. tab.$_{36|0.75\%}$

n = 36   = 30,000(0.031800)

Pmt. = $954

-258-

5. (Continued)

    (b) Total payments = \$954 × 36 = \$34,344
         Total interest = \$34,344 - \$30,000 = \$4,344

7. (a) P.V. = \$60,000   Pmt. = P.V. × Amtz. tab.$_{n|i}$

        $i = 2\%$ = \$60,000 × Amtz. tab.$_{20|2\%}$

        $n = 20$ = 60,000(0.0611567)

                                Pmt. = \$3,669.40

    (b) Total payments = \$3,669.40 × 20 = \$73,388.00
         Total interest = \$73,388 - \$60,000.00 = \$13,388

9. (a) P.V. = \$216,000 - \$36,000 = \$180,000
        $i = 2\%$
        $n = 40$

        Pmt. = P.V. × Amtz. tab.$_{n|i}$

             = \$180,000 × Amtz. tab.$_{40|2\%}$

             = 180,000(0.0365557)

        Pmt. = \$6,580.03

    (b) \$6,580.03 × 40 = \$263,201.20
        \$263,201.20 - \$180,000.00 = \$83,201.20

    (c) \$263,201.20 + \$36,000.00 = \$299,201.20

11. (a) P.V. = \$175,000 - \$25,000 = \$150,000

        $i = \dfrac{7}{12}\%$

        $n = 60$

        Pmt. = P.V. × Amtz. tab.$_{n|i}$

             = \$150,000 × Amtz. tab.$_{60|\frac{7}{12}\%}$

             = 150,000(0.01980120)

        Pmt. = \$2,970.18

11. (Continued)

(b) $2,970.18 × 60 = $178,210.80
$178,210.80 - $150,000.00 = $28,210.80

(c) $178,210.80 + $25,000.00 = $203,210.80

13. P.V. = $189,500 - $20,500 = $169,000
        i = 8%

(a) 20 Years, 8%:   30 Years, 8%:

$ 8.37 Per $1,000 $ 7.34 Per $1,000

×   169   ×   169

$1,414.53 Monthly pmt. $1,240.46 Monthly pmt.

(b) $ 1,414.53 $ 1,240.46

×   240 Months   ×   360 Months

$339,487.20 Total pmts. $ 446,565.60 Total pmts.

(c) $339,487.20 $446,565.60

+ 20,500.00 Down pmt.   + 20,500.00 Down pmt.

$359,987.20 Total cost $467,065.60 Total cost

(d) $467,065.60 - $359,987.20 = $107,078.40

15. (a) P.V. = $5,000   Pmt. = P.V. × Amtz. tab.$_{n|i}$

        i = 7%   = $5,000 × Amtz. tab.$_{5|7\%}$

        n = 5   = 5,000(0.2438907)

                                Pmt. = $1,219.45

Payment	Principal Owed	Interest (i = 7%)	Pmt. to Principal ($1,219.45 - I)
1	$5,000.00	$ 350.00	$ 869.45*
2	4,130.55	289.14	930.31
3	3,200.24	224.02	995.43
4	2,204.81	154.34	1,065.11
5	1,139.70	79.78	1,139.67
	Totals	$1,097.28	$4,999.97

*See note at the end of Problem 13a, Section 3

15. (Continued)

(b) P.V. = $6,000

$\text{Pmt.} = \text{P.V.} \times \text{Amtz. tab.}_{\overline{n}|i}$

i = 4%

$= \$6,000 \times \text{Amtz. tab.}_{\overline{6}|4\%}$

n = 6

$= 6,000(0.1907619)$

Pmt. = $1,144.57

Payment	Principal Owed	Interest (i = 4%)	Pmt. to Principal ($1,144.57 - I)
1	$6,000.00	$240.00	$ 904.57*
2	5,095.43	203.82	940.75
3	4,154.68	166.19	978.38
4	3,176.30	127.05	1,017.52
5	2,158.78	86.35	1,058.22
6	1,100.56	44.02	1,100.55
	Totals	$867.43	$5,999.99

*See note at the end of Problem 13a, Section 3

Section 5

1. (a) P = $38,000

$M = P(1 + i)^n$

i = 2.25%

$= \$38,000(1 + 2.25\%)^{100}$

n = 100

$= 38,000(9.2540463)$

$M = \$351,653.75$

(b) I = $351,653.75 - $38,000.00 = $313,653.75

3. (a) M = $30,000

$P = M(1 + i)^{-n}$

i = 4%

$= \$30,000(1 + 4\%)^{-20}$

n = 20

$= 30,000(0.4563869)$

$P = \$13,691.61$

(b) $30,000.00 - $13,691.61 = $16,308.39

5. (a) M = \$3,500,000   Pmt. = M × s.f. tab.$_{n|i}$

     i = 1.75%   = \$3,500,000 × s.f. tab.$_{80|1.75\%}$

     n = 80   = 3,500,000(0.0058209)

                        Pmt. = \$20,373.15

    (b) \$20,373.15 × 80 = \$1,629,852

    (c) \$3,500,000 - \$1,629,852 = \$1,870,148

7. (a) P = \$25,000   Pmt. = P.V. × Amtz. tab.$_{n|i}$

     i = 3%   = \$25,000 × Amtz. tab.$_{8|3\%}$

     n = 8   = 25,000(0.1424564)

                        Pmt. = \$3,561.41

    (b) \$3,561.41 × 8 = \$28,491.28

    (c) \$28,491.28 - \$25,000.00 = \$3,491.28

9. (a) Pmt. = \$100,000   M = Pmt. × Amt. ann. tab.$_{n|i}$

     i = 1.25%   = \$100,000 × Amt. ann. tab.$_{20|1.25\%}$

     n = 20   = 100,000(22.56297854)*

                        M = \$2,256,297.85

                        *Mental math was used since most
                         calculators will not show 8
                         decimal places.

    (b) \$100,000 × 20 = \$2,000,000

    (c) \$2,256,297.85 - \$2,000,000.00 = \$256,297.85

11. (a) Pmt. = \$2,500   P.V. = Pmt. × P.V. ann. tab.$_{n|i}$

     i = 0.5%   = \$2,500 × P.V. ann. tab.$_{96|0.5\%}$

     n = 96   = 2,500(76.095218)

                        P.V. = \$190,238.04

    (b) Total payments = \$2,500 × 96 = \$240,000
         Total interest = \$240,000.00 - \$190,238.04 = \$49,761.96

13. (a) M = $25,000 $P = M(1 + i)^{-n}$
      i = 0.5% $= \$25,000(1 + 0.5\%)^{-48}$
      n = 48 $= 25,000(0.7870984)$
                         $P = \$19,677.46$

   (b) $25,000 - $19,677.46 = $5,322.54

15. (a) Pmt. = $7,000 $M = \text{Pmt.} \times \text{Amt. ann tab.}_{n|i}$
      i = 1.75% $= \$7,000 \times \text{Amt. ann. tab.}_{20|1.75\%}$
      n = 20 $= 7,000(23.701611)$
                       $M = \$165,911.27$

   (b) $7,000 × 20 = $140,000
       $165,911.27 - $140,000.00 = $25,911.27

17. (a) P.V. = $128,000 $\text{Pmt.} = \text{P.V.} \times \text{Amtz. tab.}_{n|i}$
      $i = \dfrac{5}{6}\%$ $= \$128,000 \times \text{Amtz. tab.}_{36|\frac{5}{6}\%}$
      n = 36 $= 128,000(0.032267)$
                   $\text{Pmt.} = \$4,130.18$

   (b) Total payments = $4,130.18 × 36 = $148,686.48
       Total interest = $148,686.48 - $128,000.00 = $20,686.48

19. (a) Pmt. = $1,500 $\text{P.V.} = \text{Pmt.} \times \text{P.V. ann. tab.}_{n|i}$
      i = 0.5% $= \$1,500 \times \text{P.V. ann. tab.}_{24|0.5\%}$
      n = 24 $= 1,500(22.562866)$
                   $\text{P.V.} = \$33,844.30$

   (b) $1,500 × 24 = $36,000

   (c) $36,000.00 - $33,844.30 = $2,155.70

21. (a) M = $3,200,000 $\text{Pmt.} = M \times \text{s.f. tab.}_{n|i}$
      i = 3.5% $= \$3,200,000 \times \text{s.f. tab.}_{30|3.5\%}$
      n = 30 $= 3,200,000(0.0193713)$
                   $\text{Pmt.} = \$61,988.16$

21. (Continued)

    (b) $61,988.16 × 30 = $1,859,644.80

    (c) $3,200,000.00 - $1,859,644.80 = $1,340,355.20

23. (a) $P = \$1,000$ $M = P(1 + i)^n$

         $i = 1.25\%$ $= \$1,000(1 + 1.25\%)^{16}$

         $n = 16$ $= 1,000(1.219890)$

                                 $M = \$1,219,89$

    (b) $1,219.89 - $1,000.00 = $219.89